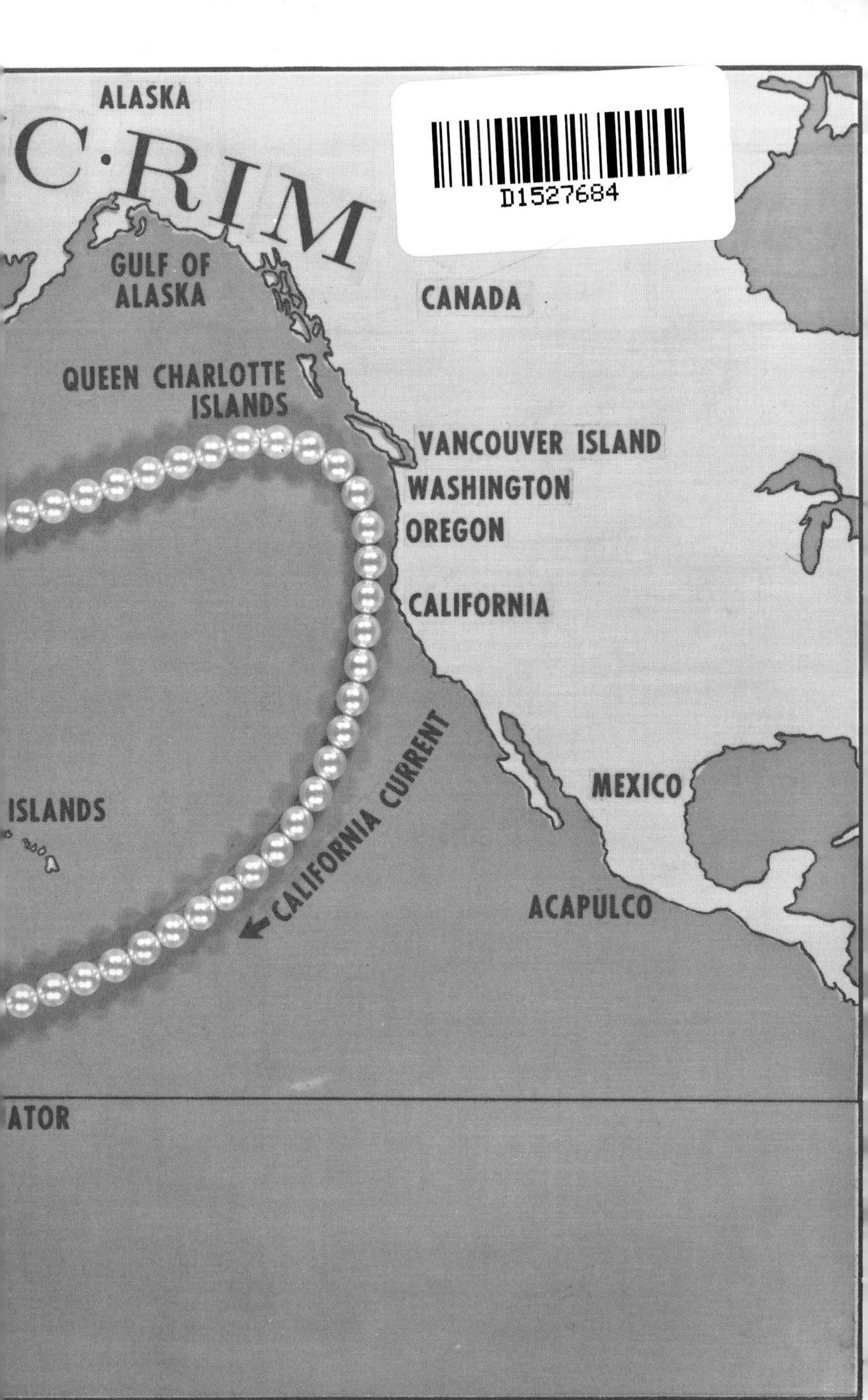
ALASKA
C·RIM
GULF OF
ALASKA
CANADA
QUEEN CHARLOTTE
ISLANDS
VANCOUVER ISLAND
WASHINGTON
OREGON
CALIFORNIA
CALIFORNIA CURRENT
MEXICO
ISLANDS
ACAPULCO
ATOR

Beachcombing
for Japanese Glass Floats

Beachcombing

for Japanese

PORTLAND · OREGON · 1971

Glass Floats

BY AMOS L. WOOD

BINFORDS & MORT, *Publishers*

Beachcombing for Japanese Glass Floats

LIBRARY OF CONGRESS CATALOG CARD NUMBER: 66-28020
ISBN: O-8323-0220-1

Printed in the United States of America

SECOND EDITION 1971

To all beachcombers
—wherever you may be

Contents

Foreword

This book is for all who have ever walked an ocean beach and speculated on objects washed up by the tides, with some such questions as — *What's this? Where did it come from? Where was it made? What was it used for?* . . . and likely a follow-up of *Is it worth carrying home?*

To me, of course, the principal prize has always been a Japanese glass fishing float, especially if it has markings on it.

Some markings are clearly Oriental; others American, Russian, or Norwegian; but some remain a complete mystery. In trying to track down what the various imprints stand for, I undertook a long search through libraries and correspondence, all with meager results. When it became evident that very little had been written on the subject, I began to research firsthand. Though this project has absorbed most of my time away from the office for a number of years, it has been an excellent excuse to explore new beaches all the way from California to Alaska.

Over a period of several years I wrote numerous letters to other beachcombers, posing questions that needed answers. I sent additional letters not only to glass manufacturing companies and related industries but also to the Japanese Trade Center, the Japanese Association of Commerce and Manufacturing, and finally even to the Emperor himself. My greatest help came from Mr. Kikunosuke Takahashi, the president of Hokuyo Glass Company of Aomori, who explained in detail the manufacture of glass floats and even forwarded a set of photographs showing their step-by-step manufacture, which I have included here.

In compiling the table of imprints and trademarks, I was immeasurably assisted by the Ralph McGoughs of Nahcotta, Washington, who likely have the largest and most representative glass float collection in the world. I am also indebted to the many friends and strangers who answered my persistent questions.

Though *Beachcombing for Japanese Glass Floats* is primarily for the beginning beachcomber, I hope that those who have already found their salt-water vagabonds will enjoy comparing their conquests with those mentioned here; and I will welcome hearing from them. Some day I should like to incorporate additional facts and experiences into the book. Writing on this subject will never actually be finished; it is just being printed at some point in time. . .

Out there in the North Pacific right now, riding the waves of the great Kuroshio Current, are hundreds of thousands of desirable floats just waiting to be driven ashore somewhere along West Coast beaches.

A. W.

Beachcombing for Japanese Glass Floats

Outgoing tide at sunset on flat Washington coastal beach. These tides carry off much beachcombing loot that the incoming tides deposit.

An Oriental glass fishing float just as it arrives on Pacific shores via the Japanese Current.

CHAPTER 1

Pleasures of Beachcombing

Among the whitened driftwood of Pacific Coast beaches can be found all sorts of interesting objects: a fruit box from Hong Kong, a hand-carved oar, a piece of whalebone, a mahogany plank—and occasionally a beautiful Japanese glass fishing float. There are treasures available to all who will search.

My wife, Elaine, was the significant figure in converting this non-beachcombing bachelor into a beachcombing husband of amateur standing. The salt-water life had been hers since childhood; but for me, a transplant from the Midwest, it was all new. Every seashell contained some mysterious life form; every blue-green float was a miracle of the sea.

We have since beachcombed most of Oregon and Washington, much of Vancouver Island and the Queen Charlottes, parts of California and Alaska—and are still moved by the excitement of finding something among the driftwood. About the greatest thrill we have during our beachcombing holidays is discovering an object we can't identify or understand: figuring out what it is, where it came from, and how it arrived at the place we found it.

Our serious beachcombing began when we visited Vancouver Island, British Columbia, in 1959. That summer, in the *Uchuk II,* we sailed out Barkley Sound toward Tofino, in search of the elusive glass floats. Tofino is a small, quiet fishing village on the outside of Vancouver Island, beyond Cox Bay.

Beachcombing this bay was a special brand of adventure. Several of the glass floats we found had a film of marine growth on their surface, and it was during the scrubbing process with powder cleanser that we first noted the Oriental markings which occur on some of the sealing buttons. From that time on, we have both been confirmed amateur beachcombers.

The beaches between Tofino and Ucluelet tell some strange stories of the sea as evidenced by the different pieces of ships and boats to be found there. In a single day we discovered pieces of an Indian dugout canoe, a skiff, and a fishing trawler; planks from an ocean-going tug, teak perhaps from a Chinese junk out of Hong Kong, and even part of an airplane horizontal stabilizer from some ill-fated event of the not too distant past. Today these smooth, flat beaches still have pilings driven in rows out into the surf, placed there early in World War II to thwart a possible Japanese air invasion from landing on that broad beach; but still standing harmless to an occasional invasion of Japanese glass floats.

There is nothing quite like combing a drift-filled beach on the outside of Vancouver Island on a sunny day; particularly during the afternoon, when the only other tracks to be seen in the sand are those of a deer, the tracks perhaps two days old; and down the beach maybe an eagle soaring and circling the slope. As the waves of the pounding surf run in over the broad, flat sand, flocks of sandpipers will run just inches ahead of the advancing water and again just inches behind the same receding wave. A crab boat that had formerly been far out may approach within two hundred yards for the lone fisherman to maneuver his Cape Islander boat to the crab-pot float, exchange pots, and be on his way again.

Armchair Beachcombing

Some of the most satisfying beachcombing may be enjoyed right at home. I have found it quite easy to start one of these

Part of an Indian dugout canoe discovered by David Close on Flores Island on the outside of Vancouver Island.

The village of Tofino on Vancouver Island is a favorite beachcombing spot. The large building in the center is the Maquinna Hotel. Near the dock, below the hotel, is a road sign reading "Western Terminus Trans-Canada Highway."

A fruit crate from Argentina, found by the author in October 1965, north of Oysterville, Washington. The box was probably dropped overboard by a ship approaching the Columbia bar. Note the blown sand about the driftwood.

Beachcombed dimension lumber stands drying outside the guest cabin at Whidbey Island. This is a single morning's haul with the outboard from the beach north toward Indian Point.

armchair hikes. For around ten dollars I am equipped with about ten charts—enough information to completely cover the living-room floor. The charts contain more data about these beaches than can be absorbed at one sitting; each viewing brings out new details.

The proper setting is before the fireplace on a snowy winter night with charts spread out on the floor and some home-canned smoked salmon to nibble on. Calipers along with tide and current tables will help the research on the particular beaches to be worked over during the next trip. Several times I have discovered a likely beach, perhaps only a narrow cove, resulting from these fireside cruises. Months and hundreds of miles later my wife and I have visited these spots and actually found floats there.

On one occasion, after studying a beach on paper for several weeks, we waited for a favorable wind and tide, then dropped everything to hike the ten miles and bring back two new trophies for the family collection. On another trip it was possible to predict very closely the number and sizes of floats that were there to be beachcombed. It is a real thrill to climb over the last rock outcropping and see for the first time an isolated sandy beach, which up to then was just a speckled area on a chart; to walk less than a few yards in the driftwood and see a glass treasure from Japan.

Practical Beachcombing

Somewhere back on my family tree, perhaps only as far as my grandfather, there must have been a strong trait of making do with what is at hand. I find it a special brand of enjoyment to roam down some beach and scrutinize every item for possible future use. For a while, my addiction was those large bolts sticking out of timbers, and it was not unusual for me to spend a whole hour rescuing some rusty-threaded carriage bolt from its former attachment. Many of these ended up in our concrete bulkhead on Mercer Island. Then came the period for dimension lumber. Much of the addition to our cabin on Whidbey Island consists of 2 by 4's and other planks foraged from the beach driftwood.

Some beachcombers carry the oddest things home, merely to cast them aside later on. I think that only those items should be kept which have immediate use or potential use the following season. But to lug items home for their artistic appeal is quite another thing. Tastes differ: a monstrosity to one person might be an inspiration to another.

Beachcombing for glass floats differs greatly from ordinary, garden-variety beachcombing. The glass-ball float is the ultimate for most beach hikers, so to go out solely for them at first is to reach for the very top. However, beachcombing for anything that suits your fancy may result in many items of interest to you personally, though these things in the eyes of even your best friends may be regarded as so much junk. But don't let that stop the search. The true beachcomber never really stops searching and often finds something of an artistic or practical nature. Imagination plays a major role in determining its use.

We walked along the Pacific Ocean beach one October day, knowing full well that several dozen people had been ahead of us. As we started searching the high-tide line, we had taken no more than a dozen steps when we chanced upon a broken oar, which now has been converted into a name sign for our Whidbey Island beach cabin, Drift Inn.

Floats from Japan

For almost fifty years, Japan, our immediate neighbor across the Pacific to the west, has unknowingly been sending good-will messengers to our Pacific beaches in the form of lost fishing-net floats. These runaways from her vast fishing industry still create as much mystery and enchantment as they did in 1918, when they first started to show up on the Oregon beaches. The mere fact that the asking and sale price for glass floats in antique shops has remained about the same is some

Paddle-end of a broken oar found at Long Beach, Washington, now serves as the name sign for the Author's Clinton Beach cabin on Whidbey.

Rows of piling driven into the flat beaches of outer Vancouver Island, during World War II—to thwart possible Japanese air invasion—still stand at Long Beach near Barkley Sound.

indication of their worth; and if a substitute is found to replace these glass spheres, then they would become even rarer and more valuable. However, to date this does not appear likely, despite the fact that some plastic floats are being manufactured.

There is a general correlation between the years of destructive tidal waves and earthquakes in Japanese waters and the unusually good years for beachcombing along the California, Oregon, and Washington beaches. Typhoons and quakes may create havoc with Japanese fishing, but they result in a bonanza for beachcombers, at least according to my Indian informant who earns part of her living by beachcombing Vancouver Island.

All sorts of things keep coming in. One year at Long Beach, Washington, there were hundreds of blue electric light bulbs. Another year, hundreds of rattan mats were found in Alaska. The travels of these items are primarily affected by current, not wind; however, local meteorological disturbances disrupt the normal current patterns and drive the voyaging items into the shore-bound surf. The determined, business-like beachcomber will look into every nook and cranny that the casual seashore stroller would pass by. Often more by imagination than by chance, he will discover some new place where an orphan float might lodge.

About the only sound we hear when hiking over the driftwood is the nearby surf slowly pounding away at the beach, but one time when we were combing a seemingly deserted beach, we heard in the near distance a racket of *whoosh, clunk, whoosh, clunk, whoosh, clunk*, as if it were following the beat of a metronome. A few yards ahead, we found the source of this mechanical noise to be upstream on a small fresh-water creek cascading down a rocky bank. Attached to some planks was a water ram, a cast-iron reciprocating device used to pump water up a pipe that disappeared over the bank to some isolated ranch house or cattle-watering tank. As we

This float, bounced up on the driftwood by the action of the surf, was easily spotted.

Author's wife, Elaine, displays Ralph McGough's mystery jug, found near Leadbetter Point, Washington, April 22, 1962. Estimated age of the jug is 200 years, but no one knows where it came from or what it was used for.

Washington State Dept. of Comme

Beachcombing for agates at Westport, Washington.

More serious agate hunters comb Westport's gravelly banks for colorful stones suitable for polishing.

hiked on and the noise died away, I wondered what the wild life that wandered by this iron monster thought about this strange noise.

How fortunate are those who live at the water's edge, where they can beachcomb at will, scanning the array of bark and kelp for chance treasures from the sea. Our occasional visits to the vacation cabin on Whidbey Island afford us this same luxury. In the early morning after a quick cup of coffee, Elaine and I have often walked the beach up to the old brick-yard plant and back. Almost always we had something to bring back to show the youngsters at breakfast. They are allowed to sleep in during these vacations, since the previous evening's strenuous games of kick-the-can or charades holds for them a higher priority than our early-morning beach walks. Later on in the day they join us for a longer trek.

All three youngsters are good fishermen and swimmers and they like to beachcomb, so we have lots of family fun at Whidbey Island, or Vancouver Island, or any other island that we happen to visit. I rather suspect that by now they have been captivated by the many moods of the sea, and, if so, that is an intended part of their heritage. Perhaps our family search for glass floats has had its more important dividends.

David T. I

Three American glass floats that were machine manufactured, as evidenced by the short necks at the top. The glass cube in the foreground is one inch on a side.

CHAPTER 2

Kinds of Glass Fishing Floats

After a ten-mile bicycle ride from his home and a full day of beachcombing, a sixteen-year-old boy stopped in for a hamburger at a restaurant in Long Beach, Washington. He had in his jacket a small float not much larger than a golf ball, which he had found in the sand at the high-tide mark. Shortly thereafter, an older couple came in and ordered steaks. In the back seat of their car, parked just outside, was a twelve-inch tan glass ball which they had purchased at a nearby curio shop.

Both the boy at the counter and the older couple at the table told about their glass fishing-float findings. The restaurant owner pointed out to the boy that his small float was a very rare color, but diplomatically did not tell the older couple that they had unknowingly purchased an imitation. . . . It often takes a practiced eye to recognize the real glass fishing floats; and like their Japanese creators, they do not all look alike.

Imitation floats are made to be sold at some—by no means all—curio stores. They are often of very thin glass and in the wide range of colors found in Christmas-tree bulbs. Some are cubic in shape, about four inches on a side. The glass is almost always free of bubbles and they usually bear a small gummed label, "Made in Japan." Oftentimes these floats are encased in string net. The one in our collection is kept to show friends the difference in appearance between these and the hand-blown floats. Some of the newer basketball-sized floats are imitations, as evidenced by the clear, colorful glass and the large silver-dollar-sized sealing buttons; but they rarely contain a trademark imprint. Apparently no self-respecting fishing-float glass factory wants its marks on imitations which wouldn't stand up long under commercial fishing usage.

For those who have never seen a glass float or heard how they are used—a typical float is about the size of a tennis ball or grapefruit and blue-green in color. For securing to the fishing net, a retaining cap is woven around it. Plastic, wood, or cork floats do not require this containment net because they have a hole through which a heavier line can be threaded. Japanese glass floats usually have a manila twine cordage in the smaller sizes and closely woven rope in the larger sizes. American-made glass floats have a characteristic tightly woven net for protection during mechanized hauling operations.

The McGough Collection

In March 1964, we drove up Long Beach Peninsula to Nahcotta, to visit Marie and Ralph McGough and to learn firsthand about their famous collection. I had heard that in just a few years they had beachcombed an amazing number of glass floats from the nearby Oysterville approach beach area, at the relatively secluded northern end of the peninsula. We found the McGoughs' namesign and drove through the ponderosa pines which lined the driveway. At the entrance to their large, older-fashioned home, we were greeted warmly.

With the aid of their small four-wheel drive truck, the McGoughs have acquired what I consider the best-organized and largest collection of Oriental floats on the West Coast—if not in the whole world. So extensive have their findings been that they have a separate building lined with shelves and racks to store floats of various sizes. They can point in any direction to beachcombed treasures which line their living room. They

Checking color of a Japanese herring net float. Note the extensive driftwood here at the north shore of Graham Island in the Queen Charlotte Islands.

Ralph McGough

Part of the McGough glass float collection, which is believed to be the largest in the world.

This float from Ralph McGough's extensive collection is one of the largest ever beach-combed. It measures fifty-eight inches around and is blue-green in color.

showed how a Japanese light bulb with a corroded base—which they had beachcombed—still gave off light when inserted into a reading lamp.

They took us for a ride out on the wide beach and even provided us with the rare experience of seeing a large dark-green float emerge out of the waves and roll up to our side of the car to be picked up. Later we held and admired dozens of their glass floats, almost any one of which could have been a museum piece. We saw their mysterious Oriental earthen jug, of which less than a dozen such have been found. Little did I realize when I found my first glass float, and wondered about it, that there could be all these variations in the vagabonds from Japan.

Mrs. McGough said that even their cat was beachcombed. She told how, during one of their family combing trips, they had found it miles up the beach, away from any habitation, in a fairly weakened condition, and had brought it home with them.

I was delighted when they offered to assist in a statistical study of floats to find how prevalent certain sizes, colors and markings were. Since they have such a wide variety within their huge collection of nearly four thousand floats, I felt the answers would be significant. Following are the results of this study:

Glass floats from the Orient appear to be manufactured in five basic shapes. To classify them simply, they are: spherical, flattened spherical, roller, cylindrical, and pear-shaped. In our collection we do have a sixth shape, a doughnut-shaped glass float, but it is American made, and experimental at that.

Spherical Floats

Spherical floats are made in many sizes, the smallest being smaller than a golf ball, the largest being twenty inches in diameter. At Long Beach, Washington, there are several floats over fifty-five inches in circumference. (In the larger sizes it is easier to measure the circumferences.)

Floats were originally made in only a few specific sizes, depending on the type of fishing that was being done then, but in later years the number of sizes has been increased to meet new requirements. The sizes found nowadays include the following (in diameter inches): 2, 2¼, 2½, 3, 3½, 4, 4¾, 6, 8, 10, 12, 13, 14, 15, 16, 18, and 19. Occasionally a flattened sphere is found—a choice collectors' item.

Roller Floats

Roller floats—so-called because they resemble rolling pins—are found as small as 4 inches long by 1½ inches in diameter and as large as 18 inches long by 6 inches in diameter. A medium size of 10 inches long by 5 inches in diameter completes the three general size categories.

The smaller roller floats appear to have been made by rolling them against a wooden form while the glass was still pliable. Some of these are tapered at the ends, almost sweet-potato in shape, and may reach to 9 inches in length. They are sealed at only one end.

The larger rollers appear to have been mold-manufactured with exceedingly clear glass, which is often pale green in color. The large ones are not often found, and very rarely for sale. The roller configuration, particularly in the larger sizes, seems to have the best built-in means of attachment of any glass float yet devised. This may well explain why so few make their way to our shores.

Cylindrical Floats

As for the true cylindrical float, the McGoughs have the only one I know of. It is 10 inches long by 4 inches in diameter. With quite thick walls, it appears to be almost unbreakable and is easily fastened to a net. Consequently very few are

One of the largest Japanese roller-type floats ever found is shown here by Chick Lovejoy of Coos Bay, Oregon.

Charles R. Hitz

Salmon gillnets with American-type glass floats attached.

apt to be found; the frequency of a find will be less than one per thousand floats.

Several sausage-shaped floats have been washed ashore in recent years. So far only one size of these has been found: 4 inches long by 2 inches in diameter. This float appears to have been hot-rolled against a form; it is sealed at one end only. A "double sausage" has also been picked up in this area.

Pear-shaped Floats

Pear-shaped floats have shown up more recently. I have seen them only in the middle sizes. They appear to be blown in a downward direction, so that the molten sphere pulls away from its stem, which is later sealed. There is considerable evidence that these are manufactured primarily for the art-store market rather than for the fisherman. Although this was not listed as a basic shape, the McGoughs found one which pulled out at each end like a football, but was only about half the size of the real thing. It had no tie grooves and was heavily constructed. For its weight, it could not have had much buoyancy; it would be considered an oddity in the pear-shaped group. And this brings us directly to the oddities of all the shapes.

Floats with Water Inside

The oddities are those floats with mistakes, flaws, or special features that make them collectors' items. We all know about that sought-after postage stamp with the picture printed upside down, which brings a price some hundreds of times its face value. So it is with floats. There are not very many genuine oddities, but being able to spot them is what separates the novice collector from the expert.

The most common among the odd floats is the one that has water inside. There are two likely theories as to how it gets there. According to one, as the nets are pulled to the lower depths, the water pressure forces the water through

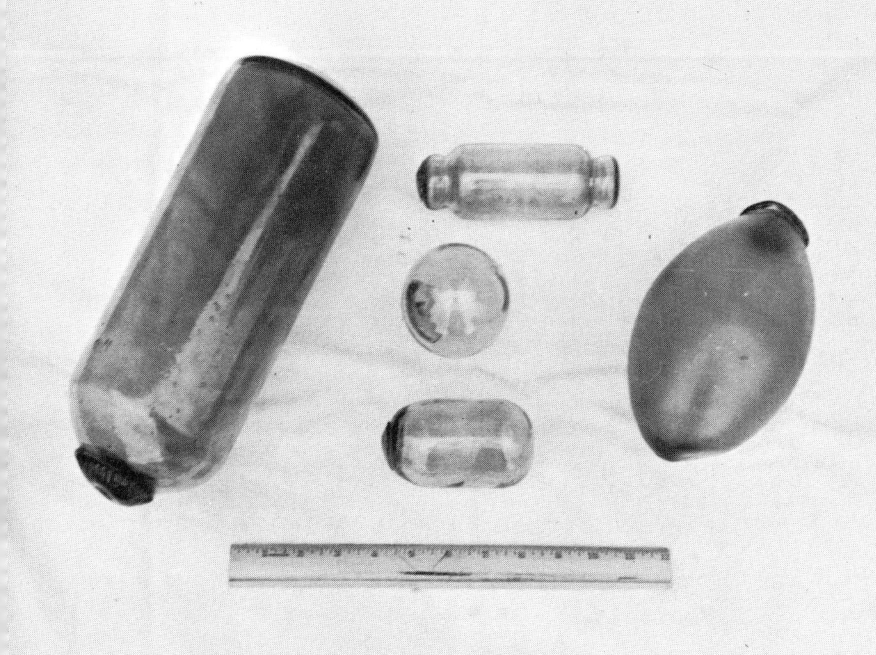

Ralph McGough

Unusual sizes and shapes of glass floats from the McGough collection at Nahcotta, Washington.

Wayne R. O'Ne

Japanese glass fishing float with extensive gooseneck barnacle growth. This ball had over thirty pounds of longneck barnacles attached when found by Ted Blodgett of Seaview, Washington.

microscopic pores around the sealed end. The other theory holds that some nets are lost to the ice in the Arctic regions, and that while in the ice the floats experience exceedingly high pressures that cause the ice to melt locally; then this same water forces itself inside the float.

There is little doubt that pressure difference has everything to do with it. However, since the sealing is done at about 1,000-degrees Fahrenheit temperature, there could be normally as little as five pounds pressure inside. With any distortion around the seal, water would leak in, particularly because most of these floats ride in the water with the seal submerged.

If a float were pulled down as far as a hundred feet below sea level, the surface pressure would be about four hundred pounds per square inch, and only a fraction of that pressure would collapse all but the most perfectly constructed float. The larger floats that have water in them are filled as much as one-fourth full, while the smaller sizes seem to have a much smaller proportion.

A story is told about a large float found at Long Beach, Washington, which contained considerable water. When the float was brought inside, it began to sing like a teakettle and tiny air bubbles rose from the seal button for two days. Another odd float is on display in Long Beach: a 14-incher about one-third filled that is reported not to have leaked in forty-seven years. Once the water gets in, it seems to want to stay.

Floats with Barnacles

Next to the float with water, the commonest oddity is the float that comes in, barnacle laden. These floats have circled in the Japanese Current long enough for barnacles, mussels, and other marine life to attach themselves first to the net and then directly to the glass. One of these floats we found on Vancouver Island took an estimated fifteen years to grow the size of sea life it carried. Some persons have kept barnacle floats like this "as is" for months. They must be kept outside though, and in this condition they are difficult to move or display because of their weight.

Floats with Spindles

The next oddity, according to rate of occurrence, is the spherical float containing an internal spindle. Apparently this spindle occurs during some unusual or accidental step of manufacture—such as an excess of glass at the right place and at the right time during the blowing or sealing operation. Chances of finding a spindle float are only about one in fifteen hundred.

The spindle consists of a filmy, internal and irregular glass tie from the sealing button straight across to the opposite side. One of these in our collection resembles a cobweb stretching from side to side. In my opinion this type of float is the most beautiful of all, surpassing even the amethyst-colored glass float generally considered the most prized by the beachcombers. Because spindle floats are so unusual, I am listing some of those known to exist:

Where and When Found	*Remarks—Diameter Inches*
Coos Bay, Oregon, 1951	Irregular stem, vertical—10″
Seaview, Washington, 1958	Blue, vertical—4″
Ocean Park, Washington, 1960	Blue, vertical—3″
Ocean Park, Washington, 1961	Irregular shape, horizontal—3″
Ocean Park, Washington, 1961	Heavy wall, vertical—5″
Ocean Park, Washington, 1962	Green, vertical—4″
St. Paul Island, Alaska, 1964	Blue-Green—3″

Double-ball Floats

Double balls—those with a small irregular sphere blown inside—are very rare. As with the spindles, these may be

A very rare glass float containing an internal filament or spindle. This glass thread extends from the side on the sealed end to the opposite inside surface. The best way to detect these spindles is by holding the float up against a bright sky as shown in the photograph.

caused by some excess of glass during the blowing or sealing operation. Double balls are definitely collectors' items. Following is a list of some of those known to exist:

Where and When Found	*Diameter Inches* *Outer Ball*	*Inner Ball*
Seaview, Washington, 1956	12″	1″ (blue)
Ocean Park, Washington, 1956	6″	1″ (white)
Lake Ozette Beach, Washington, 1960	6″	1″ (blue)
Queen Charlotte Islands, B. C., 1963	4″	1″ (white)

The Sunspot Float

Perhaps the rarest float on the coast is a milk-colored 12-incher containing a sunspot red-orange image on one side. Conjecture might explain this. The impurities within the glass appear to be concentrated locally; the spot might have been the result of the sun's working on these localized impurities for a long period of time. Or the float may have been in the vicinity of an atomic blast; in this environment the coloring might have occurred instantaneously. There is also the possibility that the float was solidly locked in an iceberg over a long period. Whatever the origin of the sunspot, this particular float is unique among eight million floats, and there may be no other like it.

Pock-marked Float

Another unusual float is a pock-marked 8-incher in the McGough collection. Most of the many pocks are on the inside, indicating that this is one that has survived a fierce rock pounding without breaking. There are a few external scars from rock and wave damage, but, with all the internal glass failures, there appears to be no single place where the glass is broken through.

During the manufacture of double-ball floats like these, the sealing button is drawn inward forming a second ball.

David T. Davis

This three-inch glass float shows an imprint or trademark on the sealing button. The frosted surface of the float indicates it has been rolled extensively on a sandy beach by the surf.

Portland Oregonian

Japanese roller-type fishing net float with unusual markings.

Frosted Floats

Occasionally floats are found with a frosted surface on the outside or inside—or both. The external frosting is probably sand abrasion caused by continuous rolling back and forth on a sandy beach. Some floats have done this with their holding nets still attached: an outline of the net is sometimes found on the surface.

Internal etching or frosting occurs when the sealing-button nubbin breaks off inside; the subsequent battering wears away the inside surface. Floats over six inches in diameter are rarely found frosted on the outside. This may mean that larger floats do not normally survive this type of beaching action, which the smaller ones can apparently take.

Other Oddities

It is said that an occasional Japanese float is found made from clear glass, free of bubbles and color, but this is hearsay only. Sometimes a large Japanese float is discovered with a glass patch added to repair a crack or bubble which showed up after manufacture. It was surprising once to see a 15-incher with a hole punctured in such a way as to show that the glass was only 1/16-inch thick, roughly the same thickness as the 3-incher. The thin section was directly opposite the seal button.

Another unusual float, on display at Long Beach, is a 10-incher with a hollow spindle or tube built into it so that a rope can be drawn directly through the float. This obviates the need for the cap net. Although this appears to be a superior way to manufacture floats, it is probably just enough more expensive to keep it from being competitive. Were these manufactured in any great quantity, I believe that—no matter how well the float might be attached to the fishing net—the general attrition of fishing gear, year in and year out, would send more of these onto our shores.

Color

Although most of the Japanese beachcombed floats are blue-green in color, examples have been found ranging all the way from clear white to deep brown. Some of the individual color tones can best be appreciated by holding the float up to sunlight. The colors come from chemicals which find their way into the glass mix, in some cases through exact control and in other cases through no control at all. Certain colors in floats are caused by the addition of oxides of iron, manganese, copper, selenium, cobalt, chromium, colloidal gold, and other agents. Addition of the oxides of manganese and selenium produce the colorless glass.

It should be noted that much of the colored glass in floats comes not by choice but from a basic mix of glass originally made for some other product line, the surplus of which is used for floats. Though blue-green is typical of most Japanese fishing floats, their floats for decorative use are often rich golds, reds, and browns. A clear pale green is commoner among commercial German floats, while American-made floats are generally clear and colorless.

Following are known colors occasionally found in beached glass floats, with the main controlling additive producing the color:

Color	*Description*	*Controlling Additive*
Clear	Seemingly colorless	Manganese dioxide
Milky white	Found in glass marbles	Tin oxide
Pale pink	As in table crystal	Colloidal gold
Light green	Found in big rollers	Iron oxide
Blue-green	Aquamarine, the color usually found in coke bottles	Iron oxide, iron salts
Green	Emerald green	Copper oxide
Light gold	As in table crystal	Sulphur

Turquoise	As in jade	Copper carbonate
Red	Blood red	Gold
Blue	Like the Blue Jay	Cobalt, copper
Wine	Very rich color	Manganese dioxide, cobalt
Purple	Amethyst, lavender	Manganese dioxide
Amber	As in rosin	Iron of alumina
Olive	Many blue-greens turn this color over a period of years	Iron of alumina
Bromine brown	Deep brown as in beer bottles	Manganese

Texture

Hand-blown floats made in the Orient are produced from an inexpensive grade of glass which by its mix and type of processing results in numerous bubbles. Saki bottles are made from such a mix. If an occasional bubble breaks through the float, it can be patched. Furthermore, the sale price of the average float is only a few cents, and the elimination of bubbles in glass is a long process. Bubbles in Oriental floats are as expected as those in Mexican-made water pitchers. However, most American glass factories operate automatic production machinery that controls the bubble problem. Some floats that are now found with a pale-blue color of regular consistency and with a relatively bubble-free appearance are likely made in Korea or China. The mold marks indicate they are manufactured in a three-piece loosely fitting mold and are designed for modest production. They incorporate a self-sealed feature marked by a kind of broken-off umbilical cord. This appears to be easier to manufacture since it eliminates the button-sealing operation.

Size Frequency in Floats

"Just what are the chances of finding a big one?" is a

question I often hear. It is unlikely for a newcomer to any beach to find a very large float right away. The first one I found was the size of a tennis ball. Ralph McGough checked his huge collection of beachcombed floats and produced the following data:

Size (diameter inches)	*Percentage of Total*
2-3	2 per cent
3-5	55
5-6	20
6-8	8
8-10	6
10-13	3
13-14	2
Over 14, below 2, and rollers	4
	100

To see if there was a variation in percentages, I made a check of our own floats found on Vancouver Island. Though the sample was much smaller—about one hundred floats—the results were very close to McGough's percentages:

Size (diameter inches)	*Percentage of Total*
2 (golf ball)	5 per cent
3 (tennis ball)	50
3½-6 (grapefruit)	25
8-12 (basketball)	15
Rollers, special, etc.	5
	100

Both locations where the floats were collected for these percentages are broad, flat sandy beaches where little damage is apt to occur to the floats during beaching, unless consider-

A large Japanese glass float containing the hemp-rope attaching net with its identification nametag still intact. Note how the size of the float dwarfs the pencil in the foreground.

A single small glass float in a Japanese lacquered bowl filled with unpolished agates. The miniature float in this picture is dwarfed by the pencil in the foreground.

able driftwood is also being washed up. And both sets of percentages answer the original question: "Just what are the chances of finding a big one?" The answer is that there is a seventy-five per cent chance that the first glass float found by the beginning beachcomber will be the size of an orange or grapefruit. Chances are even slimmer of finding a "big ball" on the rock-bound shores near Cape Alava or on the cobblestone beach of Taholah, on Washington's western coast. In both of these places, the chance that a big float will survive that final lap through the surf is almost nonexistent.

The average-size float is easily covered over with bark—or it might just roll from view into the driftwood. To find a float out in the open by itself is rather unusual, except after a violent storm.

Floats with Trademarks

"What are the chances of finding one with a marking on it?" is another question often asked. Thanks to Mr. McGough again: he checked his collection and found that ninety per cent of the floats measuring eight inches or more in diameter have markings, but that only twenty per cent of those four inches or less have them.

Rarities

The results of McGough's study of his collection, relative to the frequency or probability of finding certain oddities, colors, and so forth, have been tabulated below. These results give an excellent idea of the scarcity of some of the oddities:

Visible Cracks (several inches long)	1 in 20
Serrated Seals (10″ to 12″ diameter)	1 in 20
Rollers (4% of total) with Markings	1 in 35
Containing Colors (other than blue-green)	1 in 35
With Water Inside	1 in 150
External Sand Abrasion	1 in 300

Internal Abrasion	1 in 500
Sausage-shaped	1 in 700
Pink or Lavender in Color	1 in 800
With Internal Spindle	1 in 1,500
With Double Ball	1 in 2,000

In other words, if we took ten trips to the ocean each year and found five floats each trip, we would probably have to beachcomb until after the year 2,000 to acquire a float with an internal spindle or an extra bubble inside.

Another frequent question is: "Why all the different sizes and shapes in glass floats?" The answer is that different sizes are required for different types of fishing, and various shapes provide special attachment methods for the fishing net. As for difference in color, that has been shown to be the result of certain impurities in the initial mix, later exposure to sunlight, or some other accidental factor. . . . and apparently the fish don't care whether the net is buoyed by a float made of green cheese or purple glass.

Verna Slane Photo

A beachcomber's paradise—miles and miles of beach, beach, and more beach. This particular area is near the mouth of the Salmon River, north of Lincoln City, Oregon. You reach it by crossing the Salmon River by boat or by hiking three or four miles along the ocean cliffs.

CHAPTER 3

Techniques of Beachcombing

Glass floats can be found readily in three general well-defined areas: out in the ocean traveling in normal current patterns, on beaches where these currents impinge, and in beachcomber shops, preferably near these same beaches. Fishermen and sailors continually report seeing glass floats at sea. Ralph Widrig, a former crab fisherman out of Tofino, said that floats have been seen regularly out twenty miles or so, drifting parallel to the outer coast of Vancouver Island.

As a member of the *Sea Fever* in the Trans-Pacific Sailboat Race of 1961, Bob Withington reported seeing floats while en route to Hawaii. On the *Sea Fever*'s return, one large float picked up was so heavily laden with marine life that the attaching cord net had to be cut away before the float could be hauled aboard. Attached to the net were barnacles and mussels almost two inches long, with several small crabs hitching a ride somewhere. Again in June 1962, during the North Pacific Race, floats were seen well off the Washington Coast by members of the *Sea Fever* crew. As early as July 1934, John P. Tully of the Pacific Biological Station at Nanaimo, on southeast Vancouver Island, reported seeing many floats riding the calm sea that prevailed for a few days near Wreck Bay and the Bajo Reefs. In 1947, a fisherman out of Newport, Oregon, said:

> I got these two 15-inch floats while on a fishing boat about 125 miles off from Astoria. It is not always easy to

pick up such floats at sea, because the bow waves of the boat drive them away before you can get close enough to reach them. Sometimes it is necessary to make six or eight runs past before it is possible to get them. We see lots of floats at sea, but they seem to come in a spotty, uncertain way. Some trips we will see dozens of them, at other times not any. They appear anywhere, but sometimes they are entangled in the kelp beds. I was present when one man got a very large light-purple or amethyst-colored float. They say that these purple floats are from the nets of the Japanese Emperor. The two floats here were picked up in late August, and the bottoms were all grown over with seaweed and sea grass when I found them.

Since very few floats are plucked from the open sea because of the difficulties involved in picking them up, it would appear that a journey out to sea for glass floats is not the best way to secure them. To purchase a float in a curio shop is fine for the collector or the person who doesn't care to brave the elements, but it is like buying rainbow trout at the Pike Street Market—they never have the same flavor as those caught and then fried over a campfire. And to obtain Oriental glass fishing floats, I recommend beachcombing.

A beachcomber's future bonanza could start as it did in March 1958, off Russia's Kamchatka Peninsula:

Masayuki Katsumoto, first mate of the Kisen Sokobiki *Ami Gyosen*, of the Nichiro Gyogyo Kabushiki Kaisha Fishing Company, looked concerned. They were far at sea. The barometer was low, ominously low, lower than he had ever seen it in his eighteen years of fishing. They would have to ride out the storm; their home port of Etorofu in the Kuriles was much too far away . . .

Weeks and even years later, mute evidence of this storm continued to wash up in jigsaw fashion on Pacific beaches. Katsumoto and others of the crew were rescued by a passing trawler, but the *Ami Gyosen* had gone down in two hundred fathoms. With the violent rocking motion of the bottom cur-

Lucille McCain

Typical high-tide debris left after a storm on Washington beach, north of Columbia River north jetty. North Head Lighthouse shows at far right.

Hokuyo Ltd.

Japanese fishing boats such as this one lose thousands of floats yearly to the great "Black Stream" that is the Japanese Current.

Lucille McCain

Lost fishing floats are placed here in random fashion at an ocean-front home near Ocean Park, Washington.

rents, the fishing vessel gradually broke up. Pieces of ship and gear continued to break off and float to the surface. Portions of nets popped up, only to be further reduced in size by the action of the waves in the open sea.

In 1960, two years after the wreck, Mrs. Paul Hayes, a Clayoquot Indian living on the outside of Vancouver Island, found a large portion of one of these nets partially buried in the sand. She cut away its twenty-eight glass floats and took them to the local bakery shop for sale to occasional tourists.

The Indian lady would not have been able to engage in this simple commercial venture, had it not been for a special series of circumstances. Had any one of the following primary factors for success been absent or out of sequence, portions of Mr. Katsumoto's nets would have continued on their drifting circuit, perhaps to be bashed into some rocky Alaskan shore or rolled onto some Pacific Island beach.

Following is the series of events leading to Mrs. Hayes' discovery:

1. Japanese fisherman loses fishing gear.
2. Local currents carry floats into Kuroshio, the main eastbound North Pacific current.
3. Glass floats follow this Pacific Ocean current, relatively unaffected by wind.
4. Some floats are carried into offshore boundary layer currents.
5. Some are brought near the Pacific coastal beaches of the United States.
6. Regional Pacific Coast storms cause disruption of normal boundary-layer current flow.
7. Some floats are then carried toward land.
8. A strong onshore wind takes these floats toward the surf.
9. A high tide deposits these same floats on the beach or bashes them against rocks.
10. If the beach is relatively flat, they remain until the next high tide—unless some beachcomber meanwhile carries them off.

It is a wonder that we find any floats at all on the beaches, and those that we do find are only a small percentage of the number lost by the fishermen. Those that survive the long, hazardous journey may lodge in the driftwood of almost any sand beach along our west Pacific Coast.

A beachcombing vacation at Cox Bay (near Tofino) plus a violent storm brought us a bonanza of forty beautiful floats in March 1960. We encountered the storm on the old logging road into Tofino, where we were to be the guests of Ralph Widrig, the erstwhile crab fisherman. The tortuous sixty-five miles over which we drove in darkness and wind-blown snow, with broken branches sailing horizontally through the air, was a five-hour drive we will always remember. We had to remove trees that had fallen across the road and ford several streams.

Arriving at the cabin late and hungry in a driving rain, we viewed with little enthusiasm the prospects for a gay vacation. All that night the driftwood boomed with the roar of the surf just beyond our small cabin. Many small boats were lost in the torrential rains and high tide, and the province pile driver broke away to drift down the inlet. When morning came, the storm was still in progress and rain was still driving hard. Our host said:

"Today might be good for floats. The radio reported winds last night off Tatoosh of ninety knots, and the Lennard Island Lighthouse measured eighty-five knots here. The wind is steady but will subside. The storm is moving on, and the high tide will be about four p.m."

With that announcement, we hurried out to the nearby beach. Within minutes, we picked up two floats, barely a hundred yards from the cabin door. By dinner that night, after a search of Long Beach (between Ucluelet and Tofino) and Schooner Cove, the total count was forty, including several floats larger than basketballs. Most of these were found on the open beach, the rest within the first few feet of the driftwood. One rolled right off a breaker as it spilled out on the

Amongst the driftwood pushed by the surf thirty feet above high tide, David Close points to the place he found a glass float.

Here it is, a blue-green Japanese glass float.

flat beach, and one had mussels three inches long growing to the attached net.

High-Tide Kelper

One of the most successful collectors in this area is my friend, Oates Cochran, who says he has found the right formula. His success is based solely on the matter of timing, at least for the Washington Coast beaches. As far as his type of beachcombing is concerned, he says he is a confirmed "high-tide kelper"; he has never had any luck searching the driftwood. For the outer Washington beaches, he says you must be there at high tide and a strong west wind must be blowing. Just as the tide turns, the floats may roll right up and then roll right out again—if you are not there to retrieve them.

As for the "kelper" part, he has worked over kelp beds near a rocky inlet where many of these floats become enmeshed because of their extensive marine growth. He said at first he didn't recognize the floats because they looked like kelp bulbs, but once in a while a bald spot would betray their glass surface. He found eight floats this way at night with a flashlight and rolled them up onto the sand, away from the surf for safekeeping. Later he collected them in a potato sack to carry back to his car. He, too, beachcombed the big storm of March 1960, but from behind the north spit at the mouth of the Columbia River. His top prize from a huge pile of kelp was the mast from a sampan, a fishing skiff used in the river and harbor traffic of China and Japan.

North Sand Point Beach

It is important to know where to look for floats. In June 1961, we hiked three miles past Lake Ozette to the North Sand Point Beach, only to find several other families there, all pawing through, over, and at the edge of the beach driftwood. It looked at first as though it would be pretty slim pickings. Since it was a such a beautiful day, we decided anyway to walk down the beach. We followed the water's edge at the high-tide mark—which for that day was about 150 yards out from the drift—on the flat, sand-covered beach. The surf was modest. After a short time, and almost in front of us, a 3-incher rolled out of the surf like a baseball up toward our feet. Pretty lucky, we thought. All the others were laboriously searching the drift. In our hike to the point and back, we found four floats at the water's edge and spotted several down the beach from highlights reflecting in the bright afternoon sun. One of these was grapefruit size, another a roller pin of quite good glass.

Not always does the story turn out so well. A friend related that, during one of his trips north of Pacific Beach, Washington, beachcombing had been rather sketchy because of the early-morning patrol of the local jeep fleet that scoured miles of beach within a few minutes. After a long hike to the end of this particular beach—which was divided from a more northerly part by a sizable stream—he spotted a beauty out in the middle of the stream. The current would carry it out, then the surf would bring it back, out and back, and out and back again. Finally he gave up in disgust. The current was much too strong and deep, and the surf too ominous, for him to attempt to retrieve it. Like trout fishing, the ones you remember best are often the ones that get away.

And here is the story of one that almost got away: A neighbor lady, who has a cabin at Ocean Park, Washington, tells of watching one with her binoculars for some time while it was still in the surf. Finally it started to come in close enough for retrieving, so she put on her boots and went out into the surf, still keeping an eye on the ball. Suddenly a man in a jeep drove past her, farther out into the surf, and proceeded to go after it. During the soprano tirade that followed, she got her 10-incher. It is a real prize with four flat spots, unusual for that size.

There are risks for those beachcombers who would drive the flat, sandy beaches. I understand that in the Long Beach

Brenna Photo

The driver of this car at Ocean Park, Washington, took one chance too many—a single high tide caught him off-guard.

"Scanning a beach for floats is somewhat like looking for agates."

area about a dozen automobiles per year fall victim to the incoming tide. Frequently when a car gets caught by the water it is never salvaged. A pickup truck was lost in October 1963, when a newcomer to the beach, some five miles north of the Oysterville approach, hit a sand swale about two feet deep. The splashing water killed his engine, and he was forced to walk nine miles to Ocean Park to get the wrecker. When the water receded, all that was visible was a bit of each back tire; the rest was buried upside down. He couldn't even get the back wheels off, the sand filled in so quickly. After the next high tide everything was out of sight.

The Art of Scanning

If you don't find your float among the fishing boats, or at the water's edge, or in the kelp, then you might learn the tricks of working the driftwood, as well as all other areas at the high-tide levels. On a flat beach the high tide will leave residue over a broad area; this should be scanned first. When scanning an area, scan all of it. I found one day that I was concentrating too much on distant detection and missing items right underfoot. After this happened a second time and I was asked about that green bottle or that piece of gear which I had missed, I revised my scanning to include the "nearby" out to thirty feet; and the "faraway" out several hundred yards. Scanning takes practice but will improve markedly by the second day; by the end of the week very little gets past even the amateur.

It should be evident by now that ninety-nine per cent of the success in finding floats is in the detection process, that is, if the location and time are right. The remaining one per cent is in the actual acquisition. Detection is all important. The visual scanning skill of one person as related to that of another will probably measure the day's results.

Scanning a beach for glass floats is somewhat like looking for agates. Some persons have "agate eye." They can look ahead and immediately spot the glimmer or reflection from a glass surface; or, during a bright, sunlit day, they can look down into the shadows of piled driftwood to spot a float. Such a person is Don Close. I have walked a gravel beach straining to spot an agate, when Don has called out directions such as these:

"There is one right in front of you, fifteen feet away." I would walk up and start searching. He would then say: "It is three inches inboard of your big toe." And, sure enough, there it would be as big as a large pea.

In April 1962, at Schooner Bay, British Columbia, seven of us beachcombed a total of thirty-six man-miles and found two small floats. Don spotted the first one in deep, heavy driftwood, down five feet between two large logs. His son David found the second one in almost identical conditions. Furthermore, there was considerable evidence that this beach had been well combed the previous week by a dozen people. To me this trip proved to be the ultimate test of scanning prowess.

Don doesn't know why he has this ability. It may be the result of training his perceptive abilities to separate the dull surface of an agate from the shiny surface of wet gravel, or to spot that faint glint of glass in the dark shadows of the drift. Another friend says that he personally can readily spot agates by looking for them while walking directly into the sun: the sun's rays, going through the agate, cause it to glow. Another states that he spots agates best when the sun is shining just over his shoulders.

I find that when combing down a beach, the percentage of "find" on the first sweep will be high, something like seventy per cent of the total. On the return trip the percentage is usually low; however, the sun, shadows, and angle of search into the drift are different, so the trip back is always worth while. If total time is limited and a certain number of miles must be covered, a reasonably effective search pattern can be worked out by rapidly walking the high-tide mark down the beach,

Elaine Wood points out a glass float buried in sand after a tidal wave hit northern beaches. . .

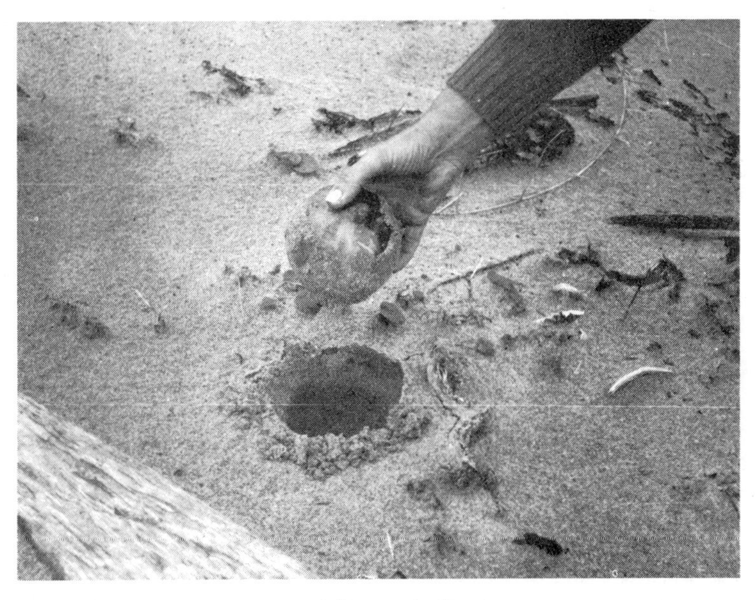

. . . and digs out the float.

This lucky beachcomber spots a small float at driftwood edge. In her left hand is part of a Japanese plastic float.

Here the beachcomber is not so lucky. She finds a large, but broken, float. It probably broke as it fell from the top of the heavy surf.

scanning left to right as you go, and zigzagging over the driftwood on the return trip.

It is still fundamental, though, that in beachcombing there should be no schedule. When you get to the point where you think you have to hurry back, your true beachcombing has already stopped.

Driftwood Combing

The heavier and thicker the driftwood is along a beach, the better are the chances, generally, of finding glass floats. For me, the mere presence of driftwood is the clear signal that all sorts of small things will be deposited on that beach by the same surf action. Since not every beach collects driftwood, I suggest avoiding the steeper beaches with a bluff. This is no place to be caught at high tide. The slope of the beach, the prevailing winds, and the coastwise current action are all factors determining how much of the flotsam will remain. As a rule, most North American beaches experience a current flow that is parallel to the shore. Were it not for the local storm and surf action, the glass floats would keep drifting past. Certain rocky points of land along the Pacific Coast may create a local current condition in the lateral flow pattern to catch some of the floats. Usually these same rocky points acknowledge the prevailing current flow with a wide, sandy, driftwood-covered beach on the lee side.

An amateur beachcomber really has to work to get results among driftwood. He must take care that each log is securely enmeshed before starting to climb. To lose footing or fall from a high log to a lower one will injure one's composure—and possibly more. Besides, that new log recently stacked by the last high tide never really warns about the thin layer of slippery green growth painted transparently on its surface. Because it is a rare day that I don't fall once or twice, I usually wear ankle shoes and thick trousers and gloves. The innocent-looking dry cedar logs with their chewed-up ends

Large logs often conceal glass floats or bottles that have been tossed there by the waves at high tide. It is easier to walk the long boom logs than the gravelly beach, but care should be taken. These youngsters have found that a walking stick helps.

Great stretches of driftwood like this mean many places for interesting things to be hidden. Zigzagging through the drift often brings good results.

provide thousands of little splinters for the amateur to take home in his fingers.

In heavy drift, one of the fine points to be learned is to avoid walking the long, easy boom poles. The reason for this is that most people working the drift will try to eliminate climbing up and down. True, by walking on the tops you may command a better over-all view, but at this point you should be concentrating on the hidden recesses. I have watched many groups start at the beginning of the drift, and invariably a great majority will take the long, high, easy trails. Working at the lower levels will be much harder and slower but more productive.

Once you have found a float hidden in the shadows of log drift, don't pick it up until you determine the angle it was viewed from, and you will be surprised at how narrow that cone of vision sometimes is. Another part of this technique of looking is to study the highlight which first caught your glance. This too will be revealing. On a wide, flat, open beach when the sun is shining, a glass bottle or float can be spotted a long distance away. In rainy weather under a thick overcast, the detection radius is quite short, even on a clean sandy beach free of drift.

I find glass floats at the same level in the driftwood where there are bottles, pieces of cork, bark, and other small pieces of wood. An occasional spent light bulb tells me I am on the right track. Some floats will roll down into nooks of the driftwood to be lost to sight, but the confirmed beachcomber will ferret them out. After hiking the logs of thick drift and peering down through all sizes of wooden planks, bark, and flotsam, I have often wondered how many hundreds of glass floats have worked their way into dark inaccessible locations. There is comfort in knowing that the next winter storm, coupled with a high tide, will release these hidden floats. Meanwhile the driftwood is being driven into a new pattern with a new crop of floats for the beachcomber to locate.

Old weather-beaten driftwood, silvery in color and sometimes fallen to rot, is generally not productive as far as glass floats are concerned. The associated moss indicates inactivity where the tides are involved, and this is not characteristic of the life and voyages of a glass float. I have yet to find a float in old drift.

One area I beachcomb is the grass and bush up beyond the drift. Technically this is known to the surveyor and map maker as the "line of vegetation." I recommend that the first ten feet behind the grass-line be searched. This won't be easy, particularly where the salal is thick; however, it is surprising how many items at high tide get driven or flipped back this distance by the wind and surf. At high tide, when logs bang one another, the small glass float can be bounced quite far back. I have found several small floats in thick grass a foot high, about fifty feet back from the edge of the drift.

Digging for Glass Floats

Most people associate the hunting of agates with walking a gravel beach and searching the gravel, but the real rock-hound and other knowledgeable persons go into the hills looking for certain rock and soil strata where agate originates. Agates the size of baseballs are sometimes dug up in the soil of certain logged-over areas in and around large stumps. Near Chehalis, Washington, there is an excellent place. Agates are detected by probing deep into the soil with iron rods. And so it is with glass floats.

To probe for floats in sand banks, a light-weight skewer is recommended, similar to those used for shish kabob barbecuing, preferably about thirty inches long. A twenty-inch skewer is about as short as you want to use; however, a skewer of almost any length is a useful implement around a beach campfire. A broken fishing rod with the eyelets removed would also double for a probe. I use a thirty-five-inch aluminum rod, one-fourth inch in diameter. It is light and not too long for my pack. It also serves as a walking stick, although a heavier stick is much better when combing driftwood.

Digging out the object located can be done with a clamshell; therefore no trowel or shovel is required for your beachcombing equipment list. When skewering the banks for floats, don't be surprised when your divining rod detects buried logs as well as fish-net balls.

In August 1961, we mined a number of glass floats out of the sand banks of Vargas Island. Eight of us in two boats made the ten-mile trip west of Tofino, and while one boat fished for salmon nearby, we beachcombed forty-eight man-miles that afternoon and brought back twenty floats. This I considered excellent for the usually nonproductive summer season. The spring and summer of that year had been quite dry for this part of Canada; consequently some of the sandy banks had slid away, exposing several floats buried as deep as one foot below the grass turf.

At one spot we dug a float from the sand at eye level in a six-foot-high bank. In another case, a float had obviously been partially exposed in a sand bank for a long time because the buried portion was still blue in color, but the exposed part had been changed to an olive color by the sun. The total find of our party might have been greater had we all been armed with steel rods to spear the sand in the immediate vicinity of the buried floats.

When to Look

If given a choice of when to beachcomb, mine would be the second day of a storm when the strongest winds are blowing across the beach. The high-tide time is important too, but in my opinion the wind factor has greater priority. Strong winds blowing parallel to the beach are about as dampening to the dreams of the float collector as finding automobile tire

Hjalmar Bren

Hjalmar Brenna of Nahcotta, Washington, shows some of his beachcombed loot and recommended garb for wet coastal weather.

tracks in the sand from a new access road which has just been bulldozed through "impossible" brush to a once inaccessible beach. Quartering winds will often deposit floats on the windward side of a sandspit or rock point.

My next choice of the best time for beachcombing would be a Thursday of the most overcast rainy week in March. Why Thursday? That is the day to avoid week-enders (the Friday through Monday visitors), and it allows three days of tide changes to accumulate all sorts of beachcombing treasures. My third choice would be the highest tide of the month, regardless of the time of day. A high tide with an onshore wind will leave floats on the beach during relatively calm surf action.

What to Wear

When tramping the sandy beaches and rock outcroppings of the Pacific Coast during spring storms, we encounter considerable rain. Much of the outer Washington and Vancouver Island area experiences as high as 140 inches of rainfall per year. For a week of spring vacation at Tofino, we plan on five inches or so of that total—which brings us to the subject of clothing.

The garb of the local fishermen gives us the major clue as to what is best for this area. For the cheechako or amateur beachcomber to be comfortable from dawn to dusk in March temperatures and rain conditions, he had better begin his wardrobe with long-john underwear, then add a wool shirt and trousers, with a fisherman's black-rubber gear over all the rest. With his feet encased in rubber lace-up boots, his hands protected by waterproof gloves, and his head crowned with a southwester rain hat and flaps, the beachcomber is ready for the elements; in fact, ready for all 140 inches. In this outfit he may not look so attractive but he will be snug and dry, well fortified for whatever the spring coastal weather has to offer.

Clothing suggested for beachcombing in August is quite another matter. For the warmer weather, bermudas, sport shirts, and a wool sweater will do nicely. Beach tennis shoes are fine for Vancouver Island, but not so fine farther south. There where it is more populous, with more lumber mills, more dimension lumber, and more boards with nails in them, I usually wear ankle shoes—the same as for driftwood combing. The youngsters in their tennis shoes slosh right through small streams while I must detour or wade. Hip boots are fine, but the nail possibility must be kept in mind. Gloves are handy for climbing the drift or beating a path through the brush. If you lose your footing at the edge of some slippery rocks—grabbing a small spruce or wild rose bush, with its half-inch thorns, can be a memorable experience.

In selecting summer clothes, don't forget that August brings different weather at different places. August at Tofino is very different from August at Nanaimo, yet they are at the same latitude on the same island. Outer coastal weather is often, and quite rapidly, subject to the whims of the Pacific anti-cyclonic wind center, with its accompanying thick overcast. Then again the weather can be clear, with sunshine for weeks.

Cowichan Bay Indian Sweaters

A wool sweater is recommended even for summer weather on outer Vancouver Island, and the local hand-woven Indian sweaters will suit any weather that comes along. The famous Cowichan Bay sweaters are made in the vicinity of some of the best beachcombing and salmon-fishing areas in the Pacific Northwest. These hand-knit, raw-wool sweaters are now seen almost everywhere in the Northwest, whereas only a few years ago they were the hallmark primarily of coastal British Columbia fishermen and outdoorsmen. They keep the wearer dry and warm and will last a lifetime.

Each of these sweaters is a memorial to a Scotswoman back in 1890, whose name is not now even remembered. She

Norm Thompson

The famous Cowichan Bay raw-wool Indian sweaters are made in the vicinity of some of the best beachcombing and salmon-fishing areas in the Pacific Northwest.

became concerned over the plight of the Indians at Vancouver Island. To provide them with an additional craft with which to increase their standard of living, she taught them her former Shetland Island skill of carding, spinning, and knitting the raw wool. The three basic natural-wool colors of gray, beige, and black were woven into original designs. The sweater sheds the rain because of the natural lanolin retained in the original untreated, undyed wool. The Indians also now knit raw wool socks and berets. A friend of mine has a pair of these socks he bought more than twenty-five years ago that still give good service. Incidentally, in pictures of the 1898 Alaskan gold rush at Dawson, I have spotted an occasional Cowichan sweater.

What to Carry

For my ten-mile beachcombing jaunts I use a Trapper Nelson back pack. It is fine for carrying the camera and lunch on the outbound trip and large enough to hold any large floats or other beachcombed items on the return trip. This type of pack can be worn over thick clothing. Should the weather turn warm, it is still comfortable to wear when I am stripped to the waist.

A lightweight camera is a *must* for beachcombing, not only to record the actual finding of floats but also to record the huge cedar stumps and other weather-beaten driftwood that become artistic additions to your photographic collection. There are ever so many items of beachcombing interest too heavy to pack out. Besides, photos of all the things we leave behind—such as old chests, boat frames, half-buried Indian dugout canoes, and large crates with Oriental characters—will inspire friends to want to take those same trips. On Vancouver Island I once found an airplane stabilizer which had washed in right after a storm. It was filled with sand but I was able to pry it free and stand it up behind the driftwood for ready identification—and photographing. That was the time I didn't bring my camera.

Portion of the outer coastline of Flores Island, a beachcomber's paradise. Note the picturesque stump defying further relocation by the sea.

A thermos of hot cocoa or hot soup seems lighter when carried in a pack. For snacks and lunch, select the menu that suits your fancy. As for me, I am a sandwich, fruit, and cheese fan. The youngsters have their own ideas. Dick usually has pilot bread and pemmican in his pack. Francie sees to it that there are pickles and olives along. Nancy tries to convert the main course into tuna fish sandwiches. If there is heavy rain, we find shelter for lunch under a makeshift plastic tarpaulin. It is surprising how cheerful a special lunch can make an otherwise desolate windblown spot.

Polaroid glasses could be useful to the beachcomber, and I wish someone would invent a pair of special glasses for looking into dark recesses to detect glass objects. A small metal mirror would be helpful to use around the dim edges of logs. One of those surplus signalling mirrors would be excellent.

I have never used a bicycle while beachcombing. Offhand it would appear to be a good means to travel farther, if just achieving a certain distance were the objective. And, usually, there is much necessary walking over logs and dry sand. If equipped with a utility basket or saddle bags, a bicycle would be fine for carrying home the loot, downwind that is; but it would make slow headway against a gale-driven rain, and the three or so miles back to civilization might be more miserable than walking. Many times the wind will turn parallel with the beach. I have encountered winds on Cox Bay that drove the rain horizontally right through the seams of my clothes. One time at Schooner Bay the dry beach sand blew so hard that I had to cover my face; there are no windbreaks on the open beach. It is always advisable to start beachcombing upwind or into the prevailing wind, so that the return trip will be with the wind to your back.

Beachcombing with a motor scooter might add more miles of travel to a very long beach, but this is not the way I would choose to spend my hard-to-come-by beachcombing days. Very few of the beaches I would select are suitable for these mechanical assistants. Those who wish to race or drive on a flat beach are free to practice this brand of fun; but those in search of the glass float will select less populated and less traveled areas. My brand of beachcombing is done without the accompaniment of exhaust fumes.

All of these beachcombing tips on how to look, where and when to look, what to wear, and what to eat are, however, no substitute for an understanding of factors governing the travels and beaching of glass floats. The best month is March. The best time is during a storm. The best winds are onshore and the best tide is a high tide, even if it is at night. Also remember that these floats won't "stay put" for long.

To learn firsthand the techniques of beachcombing, I suggest you head for the beach on any day you choose and just start wandering. You may not find a glass fish-net float, but it will be an experience you can't afford to miss. Beachcombing is not what you find, it is what you hope you will find.

Hjalmar Brenn

Behind the large float is a wooden plank covered with gooseneck barnacles, found at Ocean Park, Washington.

CHAPTER 4

The Long Sea Journey

In the past fifty or sixty years there have been some major changes in Japanese fishing techniques and fishing areas, but the Japanese continue to use their glass fish-net floats. If you were to pick up one of these floats on a Northwest beach and it did not contain a trademark, it might have been lost anytime from 1910 to last year. If it did contain a trademark, it still might be difficult to trace it to any ten-year calendar period.

The cost of these floats is low and their efficiency high. And the Japanese should be good judges of this because in fishing they top all other nations of the world. More people are engaged in fishing in Japan in proportion to its population than in any other country. This fishing falls into two types, coastal and deep-sea, and it is rigidly restricted by the government. The Japanese fisherman must get a license which specifies in what area he can fish and what kind of fish he can catch.

Prior to World War II, the Japanese developed to a high degree the floating cannery and whale-oil factory in which the catch of fish, crabs, or whale could be processed at once without losing fishing time. So effective were these modern methods that whales became more and more scarce, and some Americans feared for the annual salmon run. Floating fish factories and the sea-going trawlers also provided the Japanese Navy with vital navigational information about the Aleu-

Barnacle mass attached to Japanese long-line fishing float picked up in Queen Charlotte Sound.

Phil Burton risks driving along the sandy stretches of Graham Island in the Queen Charlottes on some of his beachcombing ventures.

tians and the Alaskan Coast. It is believed that many of these fishing boats operating off California and Mexico performed similar services for the Imperial Navy.

War brought a curtailment to these operations but with the end of the occupation in 1952, the Japanese again started high-seas fishing on a large scale. However, as their fleets expanded into foreign waters, they found closed areas where formerly they had fished extensively. Before the war they had fished in Chinese and Korean waters, off the Kamchatka Peninsula, and in the Sea of Okhotsk. But the new limitations merely accelerated construction of larger vessels for longer-range operations, since the homeland waters were already overcrowded. This step required bases abroad, which resulted in the formation of overseas joint-fishing companies, particularly in Latin America and Asia.

The Japanese furnished technical direction, vessels, and gear. Sometimes the country involved would set up processing machinery, or again it might allow fishing rights in its waters on a share basis. It would appear that the Japanese were adept in meeting the desires and requirements of each different area. By 1958, six years after the end of the occupation, they had active agreements with fifteen countries, and twenty-three additional fishery projects were under way.

More than two hundred Japanese vessels were involved in these overseas operations, three fourths of which were long-line tuna ships. Japanese vessels were based in Argentina, Australia, Brazil, Burma, Cambodia, Ceylon, Chile, Colombia, Cuba, Haiti, India, Indonesia, Iran, Israel, Italy, Malaya, Morocco, New Hebrides, Pakistan, Panama, the Philippines, the Ryukyu Islands, Samoa, Sarawak, Singapore, Formosa, Thailand, Trinidad, Venezuela and Viet Nam.

All these added fishing waters meant that now the large long-line tuna glass floats would wash up on shores unrelated to the Pacific Ocean. John S. Robinson, a lawyer of Olympia, Washington, experienced an example of this. He wrote:

About a year ago, when I was staying with some British friends at their farm on the slopes of Mt. Kilimanjaro in the heart of East Africa, I was amazed to see on their veranda a Japanese net float—one of the ordinary blue-green kind.

They told me they had picked it out of the Indian Ocean while on holiday at one of the Kenya beaches and that the floats frequently turn up there.

The Emperor of Japan may not have any nets, but his subjects have plenty, and apparently the sun never sets on them.

Routes

As to the various routes a float may take between being lost and being found, this is even more complex. I once was naive enough to take current charts of the Pacific Ocean and exercise on them this seemingly simple problem: If a float were lost at a certain point, say off the coast of Japan, what were the routes it could take and where might it be beached? And, if it were lost at some point fifty miles south of the first point, what might happen to it? Possible answers got so mixed up in such a short time that to state beyond the most general of conjectures would be mere folly.

An experiment was conducted in July 1963, when the Citizen Trading Company of Tokyo released 120 "friendship buoys," each of which contained letters and a watch. The first buoy was found in April 1966, by Mrs. W. H. Siegfried, while she was beachcombing on Oregon's Nehalem Spit, the very beach where sailing vessels from the Orient—also riding the Kuroshio Current—foundered almost two hundred years ago.

The buoy had traveled about 4,400 statute miles in 1,000 days, so the average drift was about 4.4 miles per day. Off Yokohama, the current is ten miles per day. In the mid-Pacific it is about five miles per day, while off the Oregon Coast it is three miles per day. Therefore it appears that this first "friendship buoy" made the trip across in record time and was

Tillamook Headlight-Herald

A "Friendship Buoy" shown by Mrs. W. H. Siegfried of Nehalem, Oregon. This was the first to be picked up from around a hundred released off the coast of Japan in 1963. During its three years afloat, the brass screws holding the cover plate were almost completely corroded away. Inside the buoy was a letter of instruction, also a letter from a high school student.

Lucille McCain

Successful beachcomber with Oriental float discovered near Ocean Park, Washington.

beached without much coastwise travel. This matches reports of bottles with notes from Japan being found on Oregon and Washington beaches after about three years' travel-time en route. It is quite possible that one of the "friendship buoys" might continue in the mainstream current for another complete circuit of 15,000 miles and then show up on Nehalem Spit in the winter of 1976. Based on experience elsewhere, it is probable that no more than six of the original number may be found by 1970. So, what happens to the other 114? Only Kuroshio, the Japanese Current, can tell.

Once an Oriental glass float becomes a vagabond of the Pacific Ocean, it is subject to current and storm patterns of a body of water covering almost half of the earth's surface. Our primary interest here, though, is in the currents north of the equator.

Kuroshio—the Japanese Current

There are other currents and drifts in the North Pacific that toss beachcombing treasures ashore on our western coast, but the warm, tropical waters of the Japanese Current do most of it, particularly in the matter of glass floats. The Japanese call it the *Kuroshio*, or "Black Stream," because of the dark color of its water.

Kuroshio's permanent circuit starts as a northern branch of the west-flowing equatorial current; it travels along the coast of Japan, where it meets the arctic waters of the Oyashio Current. Kuroshio then loops east across the North Pacific, widening and slowing down as it fans out in the Gulf of Alaska, where it drops a quantity of flotsam, then heads south dispersing its treasure along the coasts of British Columbia, Washington, and Oregon.

Outside California and Mexico, it again joins the equatorial current to run westward, without any obstacles, in the longest ocean current known to man—some nine thousand miles from Panama to the Philippines, where it completes one circuit and endlessly starts another.

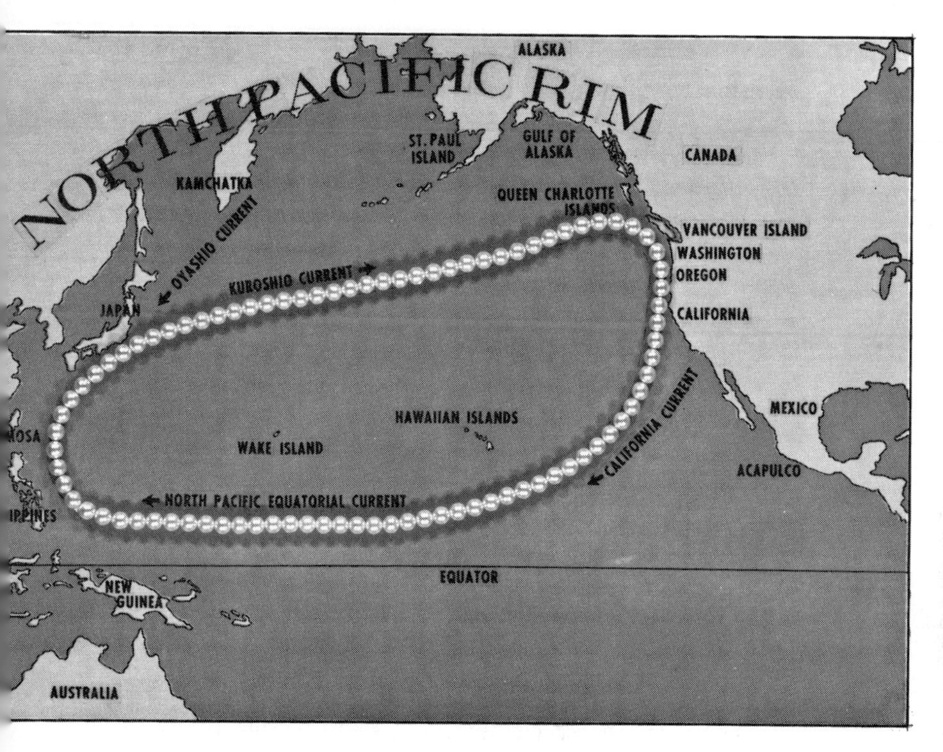

ROUTE OF THE JAPANESE CURRENT

"Roughly, forty per cent of all Oriental glass fishing floats ever lost are probably still out there in the Pacific, continuously traveling the gigantic Kuroshio circuit, awaiting the chance storm that will drive them ashore on our West Coast. Part of this mammoth necklace of glass floats has been seen off Vancouver Island.

"Assuming this necklace to be thirty miles wide and all the floats spaced fairly equally apart, then each is spaced about one-quarter beachcombing-mile from another, which generally matches my findings of several floats per mile."

Here in Queen Charlotte Sound, off British Columbia, Charles Hitz has just brought aboard a large float on which thousands of gooseneck barnacles are hitching a ride somewhere.

Since most of the Japanese fishing fleets operate in the Pacific Ocean, it is readily seen that glass floats lost north of the equator will be carried east in the warm currents of Kuroshio. Though the California Current pulls some floats south as far as the 15th parallel, in the vicinity of Acapulco, others do not take this southerly route. As they approach North America, they turn left toward the Queen Charlotte Islands, and northwest past the Alaska Chain on the Alaskan Current, joining the cold currents from the Sea of Okhotsk to collide with the North Pacific Drift heading east.

Floats lost from fishing operations south of the equator would be sent along multiple paths too numerous to list here. During World War II, Japanese glass floats were reported seen throughout many parts of the South Pacific by members of our armed forces.

In the long sea journey that a single vagabond glass ball takes, it rides up and down millions of waves and is continuously guided by succeeding ocean currents which direct the destiny of all that floats upon its surface. Whether we are standing on a high cliff watching an angry surf, beachcombing at water's edge, or voyaging in a ship in mid-ocean—we can neither fully see nor experience any of these current patterns which deliver a glass float lost from Masayuki Katsumoto's fishing net off Kamchatka, all the way to the flat sand spit of Nehalem, Oregon. All we can view is the upper surface and the outer edge of these currents.

I have never seen the Pacific in a dead calm, but I have been to the outside of Vancouver Island and have seen the waters quiet enough for a canoe to have been launched in them. Yet from that same beach during a storm, logs four feet in diameter and forty feet long come cartwheeling in through the surf, end-over-end across huge rocks that border the bay.

There is the story of a glass float which was deprived of an additional 6,000-mile journey—to Norway. We had invited

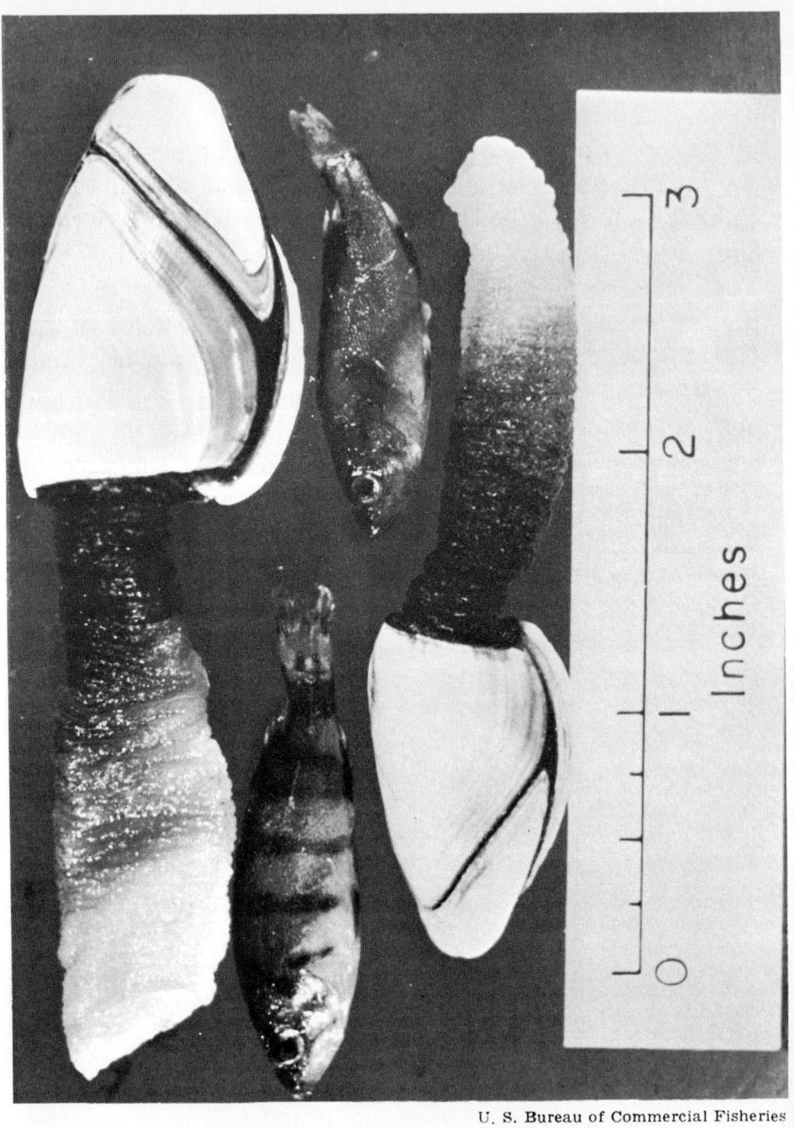

U. S. Bureau of Commercial Fisheries

These young rockfish were found living among the gooseneck barnacles on a Japanese glass fishing float found off British Columbia.

our friends, Captain and Mrs. Qvale, and their guests in for a Sunday afternoon chat and coffee. The guests, Mr. and Mrs. Dahl, were touring the United States, their first visit here from Trondheim, Norway. Mrs. Dahl related that during their previous week's visit at Seaside, Oregon, she had been made a present of a Japanese glass float as a unique memento of the Oregon beach area. Mrs. Dahl then reached into her huge handbag and, with a deep bow, presented me with the gift float saying, "We find these all the time in Norway; they wash up in our own fiord. The Norwegians use glass fishing floats extensively. Please add this to your collection as we are already overweight for the jet trip back."

And so another float joined our collection.

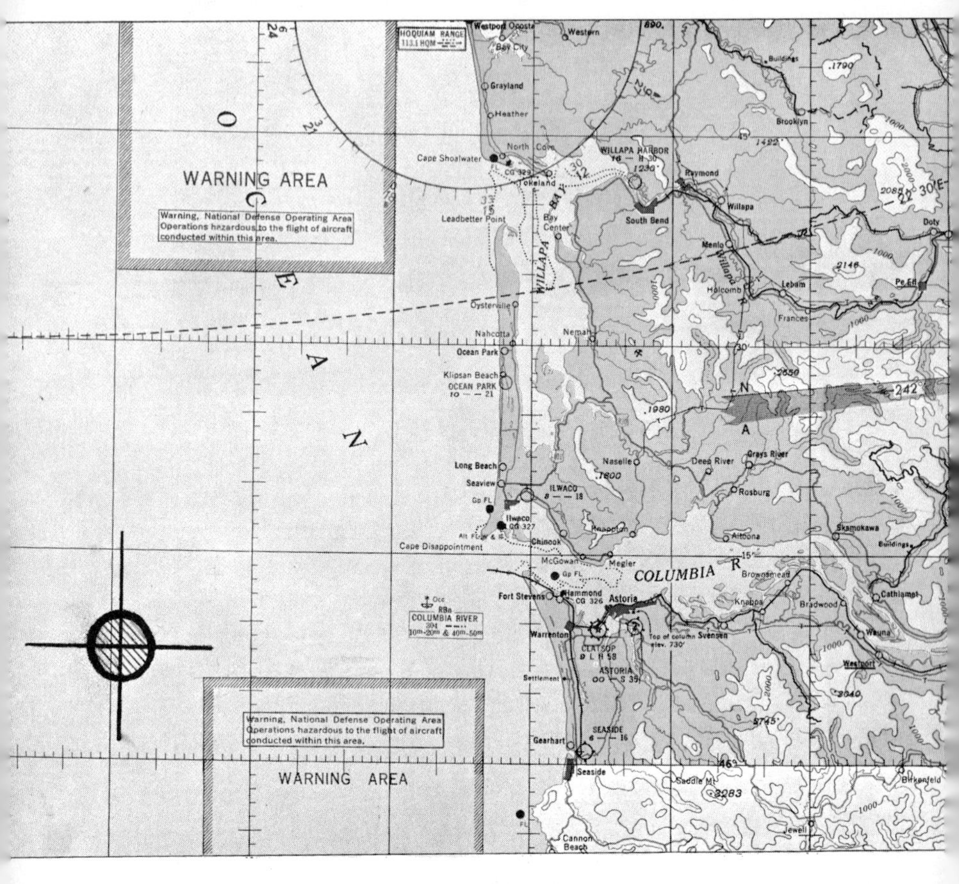

"THE OREGON MAELSTROM"

A maelstrom area with a range of some twenty-five acres is believed to occur off the mouth of the Columbia River at the approximate location shown here. Fishermen have reported all sorts of drift concentrated in the vicinity, including many large glass fishing floats.

CHAPTER 5

The Oregon Maelstrom

Various eyewitness reports tend to confirm the theory that a maelstrom — similar in nature to the celebrated Atlantic Maelstrom off the west coast of Norway—exists in the Pacific Ocean off the mouth of the Columbia River. The phenomenon is somewhat legendary. Oft-told tales describe a Sargasso-Sea type of collection in which all sorts of drift concentrate in a swirling, floating mass. The Oregon Maelstrom has been reported by fishermen as being at times as close to shore as twenty-five miles, and as far out as forty miles.

This giant whirlpool probably forms at the junction of the east-bound North Pacific Drift and the seaward extension of the west-bound Columbia River. When these two mighty forces collide where the North Pacific Drift is splitting in two —one part going north and the other part, the main part, going south—there is little doubt that at least one major vortex exists. The existence, size, and location of the Oregon Maelstrom probably vary with the seasons and the weather.

Floating masses of drift, logs, timber, kelp, pieces of ships, and so forth have been reported to cover an area up to twenty-five acres. Included in this mass are chips, and other items that float down the mighty Columbia River. Skippers of fishing boats give this area a wide berth for fear of getting some of the flotsam fouled in the screws. It is reported that within this mass there are many Japanese fishing floats. One fisherman is reported to have nosed his sixty-five-foot trawler into the edge of it, hoping to get near some of these floats—but to have quickly backed away after a sampling of the powerful currents.

Evidence of this maelstrom is scattered over parts of the Washington and Oregon beaches after severe storms. A great amount of kelp and debris such as teak planks has probably been accumulating for some time in this holding area. When a violent storm builds up, the huge floating mass breaks away from the maelstrom area and spreads out among the offshore currents. Strong storms may even temporarily disrupt the whirlpool action. If the wind is southwest—better yet, westerly—all the glass floats which have been contained in the maelstrom are blown toward shore at a speed even faster than most of the other drift. The greater freeboard of the floats causes swifter sail action than that of the more submerged items.

Max Wilson, a charter-boat operator out of Ilwaco, Washington, has proposed an expedition to search the drift, once the maelstrom is located and weather permits. In a season when the waters are relatively calm, this might be great adventure, particularly if the mass had first been spotted and located by airplane. Large, prime floats could be harvested without having to come through the surf, where they are endangered by outcroppings of rocks.

Such an Oregon Maelstrom for the collection, consolidation, and storage of glass floats—later to be discharged and distributed along the nearby beaches—seems to explain the nature and frequency of float findings on the Washington and Oregon coasts. Its action is similar to the foundering of a lumber-laden ship. The lumber is beached in a relatively well-defined area, at least within bounds to make it easy for the marine insurance underwriter surveyor to locate the hidden salvaged lumber.

It is a stroke of fortune for the glass-float collector that the beaches near this maelstrom are broad, flat, and sandy.

If the outer beaches of Oregon and Washington were rock-bound, rather than flat and sandy, far fewer Japanese glass floats would appear as decorations in Northwest homes today.

The Oregon Maelstrom would also explain why more floats are picked up on Oregon and Washington beaches than on those of Northern California and British Columbia. It is known that the North Pacific Drift current moves north and south according to a seasonal schedule, and that there is a similar seasonal movement of the North Pacific High, the predominant meteorological cell of the northern hemisphere. Knowing these general movements, we can determine the places of greatest drift-material discharge.

As Kuroshio, for example, approaches the Pacific Northwest Coast, a divergence into northbound and southbound flow is believed to take place somewhere between 45° and 48° North Latitude, depending on the season. This means that LaPush, Washington, would be opposite the northern dividing latitude; while Oretown, Oregon, would be opposite the southernmost divergence. The median latitude of Kuroshio division is very near where the Columbia River meets the Pacific Ocean. All these effects combine to make the region near Long Beach, Washington, the area where drift material should come ashore with the greatest frequency—and actual experience seems to bear this out.

The drift for the Oregon Maelstrom originates along the shores of Asia and North America and across the broad expanse of the North Pacific Ocean. From time to time, it converges into a tumultuous, revolving agglomeration centered at about 124°44′ West Longitude and 46°8′ North Latitude.

Velella

With all this complex origin and routing of Japanese glass floats, it is interesting that their arrival at the American and Canadian beaches is announced loud and clear by a two-inch jellyfish. One of the sure Indian signs that glass floats are on their way in is when Velella is washed up on the beach. Velella is a small, blue, triangular jellyfish with a sail on top. It closely resembles the Portuguese Man-of-War but is smaller and found only in Pacific waters. For some unexplained reason, this bit of ocean life, found well out at sea, is forced in by the same current and wind conditions which bring the floats ashore, except that Velella shows up on the beach perhaps a whole day ahead of the floats. When the little animals are washed up and deposited, they look like blue-colored jelly and are quite slippery; in fact they are difficult and dangerous to walk upon. Sometimes they cover a fair-sized area; we found such an area on Vancouver Island once, and the beachcombing was excellent.

So, when the Velella come in, watch the waves—and watch your step! The larger floats will arrive first, then the smaller, grapefruit-to-orange sizes. Last to arrive are the rollers, which seem to lag, and properly so, because by their shape they would have the most drag in the water.

Rogue Waves

Rogue or "sneaker" waves can be dangerous not only to fishermen at sea but also to the beachcomber. These waves probably originate with volcanic and other underwater disturbances from anywhere around the Pacific Ocean Rim. They are difficult to track and can get up to tremendous speeds.

Since 1596, Japan has had fifteen major destructive tidal waves. In 1707, one thousand ships were swamped in Osaka Bay. In 1896, a Japanese tsunami (tidal wave) and earthquake at Sanriki killed twenty-seven thousand people and swept ten thousand homes into the ocean. In the Pacific, an ordinary sea wave is rarely more than one thousand feet from crest to crest, but the tsunami is often more than one hundred miles from crest to crest. Ordinary sea-driven waves will never travel beyond sixty miles per hour, but a tsunami will run as high as five hundred miles. These giant waves are especially

These barnacle-festooned bottles and glass fishing floats were found on the beach off Salishan properties near Gleneden Beach, Oregon. In the center are five Velella picked up nearby. When these little animals arrive, glass fishing floats are on their way in. The Velella were blue in color, with a white sail located diagonally on top.

Hjalmar Brenna

Salvaging a van from the barge *Columbia*, sunk in a storm, February 27, 1965, off the mouth of the Columbia River.

dangerous on flat shores, where they may reach twenty to sixty feet in height. When entering V-shaped inlets, they can build to mountainous proportions.

A tsunami is not a single wave but rather a series of waves arriving some fifteen minutes to an hour apart. These waves may keep coming in for hours but the third to the eighth are the largest. Their onrush and retreat are accompanied by much hissing, roaring, and rattling. During one 1933 tsunami, the sea glowed brilliantly at night from the stimulation of myriads of little sea organisms known as Noctiluca. Japanese fishermen have found sardines with swollen stomachs from swallowing too many bottom-living diatoms, raised to the surface by the tsunami. In 1923, a tsunami in Sagami Bay brought to the surface and battered to death huge numbers of fishes that normally live at a depth of three thousand feet.

A dangerous aspect of these rogue waves is that, in an open sea, they may come from any direction and be completely independent of the regular wave pattern.

For the beachcomber at low tide, a rogue wave can be terrifying. A friend of mine, in January 1961, near Queets, Washington, was looking for agates in the gravel at low tide on this flat beach. Without any warning he heard a roar and saw a great wave beginning to break. With that one quick look, he started running. Just as the wave caught up with him, he was bowled over by a log which was being carried inshore in front of the crest. As he was knocked over, his feet went up because of the trapped air in his hip boots, and his head went under. For what seemed like a long, long time, he was driven right up the beach near the driftwood, head under and upside down. Then, as the wave started back, he tried to dig his fingers into the gravel to keep from being dragged seaward in the undertow. Finally he managed to catch hold on the beach floor. The wave retreated, leaving him gasping for air and in a much more precarious position, because by now his boots had so much water in them that he could hardly stand up with the added weight of the water. Fortunately there was not another wave following the big one, and he was able to sit down and remove his boots. He is now a firm advocate of loose-fitting boots for combing the edge of the sea.

I combed the northern beaches of British Columbia's Queen Charlotte Islands a few weeks after a tidal wave hit the area and found the driftwood pushed back and flattened out as if spread on the sand with a giant butter knife. It was at that level of the beach that I found several glass floats with just a small round portion betraying their hiding place in the sand. There were also many scallop and abalone shells—an unexpected result of this same tidal wave.

Fishing glass float fixed with net and rope, and fishing glass float without net.

CHAPTER 6

How Glass Floats Are Manufactured

My efforts in trying to gather factual information on how glass floats are made had been disappointing. I had written many letters but it seemed that my questions couldn't penetrate the Oriental Curtain. I soon learned that the language barrier accounted for part of the problem. I next suspected that there was also a social-class barrier, in that those people who knew most about the manufacturing did not answer letters, and those in various Tokyo headquarters offices who politely answered my letters were too far away from the problem to discuss it. I had written not only to glass manufacturing companies but to the Japanese Trade Center, the Japanese Association of Commerce and Manufacturing, private individuals, and finally even to the Emperor himself. Apparently I couldn't be successful without some sort of an agent or go-between.

I did, though, have an eyewitness of the manufacturing process. Neal M. Carter, director of the Pacific Fisheries Experimental Station, Fisheries Research Board of Canada, in 1948 had written: "During a Canadian government mission to Japan in June to December 1946, I had occasion to travel around Japan quite a bit and I incidentally happened to visit one of the plants where these glass floats are manufactured. The plant was more or less an ordinary glass-manufacturing plant, which among its other activities produced these floats, blown by hand and sealed with a blob of molten glass."

Also a friend—en route to Washington, D. C., from the Orient—called to report that, in Okinawa, glass floats were made as "cottage industry," meaning a small back-yard, oil-fired furnace with a pile of broken glass nearby. Pop bottles, parts of automobile stop lights, and so forth, found their way onto the pile.

I myself had collected a number of stories about their manufacture, and though many of them may have been true, I had no way of sorting out truth from fantasy. I was told, for example, that:

Floats are hand-blown in the streets of Tokyo by vendors.

They are hand-blown on mother-fishing-ships at sea.

They come from back-yard industry.

The techniques of manufacture pass from father to son.

Fishermen cut loose the balls because they are too bulky on deck.

Roller floats are used to buoy tins of opium in illegal Hong Kong harbor narcotic traffic.

Floats are made near the docks of fishing towns.

In the manufacture, seaweed is burned for the potash.

Floats containing marks resembling a cross had been blessed by the Pope.

Some are purposely made to contain water to be used in Buddhist religious rites.

Some floats are made by reheating amber-colored beer bottles.

Purple-colored floats come from the Emperor's fishing fleet.

Floats come from the volcano on Mt. Fujiyama.

One glass-ball factory disappeared into a crevasse during an earthquake in the 1930s, spilling its products into the sea.

No more glass floats are being made by the Japanese.

All fishing floats are now plastic.

Be all this as it may, it was obvious that I needed facts. Those facts came unexpectedly and in volume. At a Friday-evening dinner party, I was explaining my problem to my dining companion. She listened intently, asked many questions, then urged me to contact a friend of hers in Japan. This friend was an elderly lady of some circumstance and influence, who, as the widow of a tea merchant living near a fishing village, might be of some help. The following week I half-heartedly wrote the letter, asking the same questions I had written so often before. I could almost type them out in my sleep, and I knew what the answers would be. How was I to know that this time I couldn't have been more wrong?

What magic this lady used I will never know, but, in a short time, I received photographs taken inside a Japanese glass-ball factory. Within days I received other information from a glass factory in northern Japan, a source I had written to sometime before but had completely given up. A third source produced more photographs. It was as if the elderly Japanese lady had triggered the whole thing. The solution to my problem was at hand.

Photographs of glass-float production in its various stages told a great deal of the story, but there was other information needed to complete this fascinating story. I found out, for example, that the wide variation in size requirements was due to a variety of fishing methods and techniques used by the fishermen. The largest floats — having the highest buoyancy — were needed for tuna long-line fishing, in which a single line was stretched for many miles. Medium-sized floats were required for bottom tangle-net operations. The smaller and most commonly found floats were used in salmon gillnet fishing.

Of singular importance is the attrition rate of this type of fishing equipment, which is reported to average fifty per cent a year. In other words, the average Japanese fishing boat loses half its gear every year. Therefore the production of glass floats is a continuous operation in order to meet the losses of the present vast Japanese fishing fleets. Where so many nets and floats are lost, the floats must be not only efficient but they must be inexpensive. This explains, in part, the reason why cork has not been used by the Japanese. For example, I understand that today in Seattle, a salmon gillnet cork float costs about fifty cents; the equivalent American plastic float, forty-three cents, and the American-made glass float, thirty-six cents. The recently quoted price of an equivalent Japanese glass float is twelve cents.

Other elements, of course, must be considered to make the above comparison wholly factual; for instance, the American plastic float is readily compatible with the mechanized winch block, while glass floats are not. Nevertheless the extensive use of Japanese glass floats is such that many factories are engaged in their manufacture. Before World War II, approximately ten factories specialized in this; in 1948 there were four; today there are six.

One small factory that produces hand-blown Japanese fishing floats covers approximately twenty thousand square feet and employs about fifty people. The Hokuyo Glass Manufacturing Company produces the floats in conjunction with other work. A small brochure put out by the company, entitled "Glass Balls of Aomori," contains a beautiful colored selection of ornamental glass floats, glass light fixtures, and glass vases, all of high-quality colored glass. The brochure also contains an impressive aerial view of the factory, whose manufacturing operations are far more involved than the cottage-type industry rumored to be in existence. In fact, Mr. Kikunosuke Takahashi, the President of Hokuyo Glass, forwarded not only an outline of their processing but also a comprehensive set of photographs showing the step-by-step manufacture of large glass floats:

Aerial view of the head office and factory of the Hokuyo Glass Manufacturing Company at Aomori City, which produces glass floats in conjunction with other work.

Above: The trademark shown on the separate button is the symbol of the Hokuyo Company, and the most widespread marking found on Pacific Rim beaches. *Right*: Step No. 1 in the manufacture of large floats at Hokuyo: Conveying glass materials to fire-brick furnace A.

(2) Casting glass materials into fire-brick furnace A.

(3) Fire-brick furnace A (1,500-1,600 degrees C).

(4) Workmen tending furnace A.

(5) Winding up melted glass, using blowing pipe.

(6) Blowing and molding process No. 1.

(7) Blowing and molding process No. 2.

(8) Blowing and molding process No. 3.

(9) Taking off the pipe from hot glass floats (600 degrees C).

(10) Stopping up the hole on hot glass floats.

(11) Hot glass floats just stopped up.

(12) Stamping the trademark on hot glass floats.

(13) Sending hot glass floats to fire-brick furnace B for reducing fever.

(14) Putting hot glass floats into fire-brick furnace B (590 degrees C).

(15) Taking out fever-reduced glass floats from fire-brick furnace B for inspection and packing.

Packed glass floats for inland use.

Ready to start for station by Hokuyo Glass's car.

Arrival at fishing grounds, using permanent fixed nets.

Starting to catch large quantities of Yellowtail.

The current annual Japanese production of glass floats is estimated at two million units. Some of these are now apparently machine made—much like the American machine-made floats which were produced in great quantities in the 1940s. A number of these Oriental floats have been picked up, especially in the smaller sizes. They can be identified by the mold marks on the outside surface. Because few floats larger than about ten inches in diameter have been found bearing these marks, it is presumed that most of the spherical floats above that size are hand blown. Certain large rollers may be machine made, but the probable limited production and use (and loss by fishermen) would indicate that they were manufactured from wetted-down wooden molds rather than metal forms.

The mold mark of a small percentage of light-blue glass floats found along Pacific shores would indicate them to be machine made—perhaps from Russia, China, or Czechoslovakia—but no confirmation of their origin of manufacture is available.

The vast number of glass floats being picked up by beachcombers, however, gives little indication that many Oriental ones are machine made. By far the most commonly found are the hand-blown, separately sealed floats. Of course the sale, use, loss, and travel time of an Oriental float will add up to a minimum of about three years, so the beachcomber is not getting the latest model. In fact, it is likely that the average glass float picked up on a typical American or Canadian beach is easily ten or so years old.

Many of the Oriental markings are as readily found today as in the 1940s. Paradoxically, by the time you find your float, the company that made it may have gone out of business or have switched to other product lines. This adds to the overall mystery of these vagabonds of the sea and also provides the discriminating collector with additional challenges.

American-made glass floats, however, do show up regularly, particularly along the Alaska and British Columbia

American machine-made glass fishing floats in production at Northwestern Glass Company of Seattle, Washington.

As machine-made floats move along the production line from right to left, the upper neck is heated for sealing.

Machine-made floats leave the oven and are inspected prior to packaging.

shores. On Washington beaches they are found in the ratio of one per hundred. The principal manufacturer of these is the Northwestern Glass Company in Seattle. Mr. E. S. Campbell, president of this company, here explains the history of the use of glass balls for fishing floats and the development of the machine-made glass float:

> When this company started operations in 1932, it came to our attention that glass fishing floats were being used on this coast and apparently all of them were the hand-made green glass balls made in Germany, which apparently was the only source of supply at that time. We had men at our plant capable of duplicating the German-made floats so we decided to manufacture them here, and did so by the hand process for several years commencing about 1933. However, the volume then needed in this locality was small (not over ten or fifteen thousand per year), and we finally discontinued making them as our manufacturing facilities became more and more in demand for other products.
>
> In 1942, this subject was brought to our attention again as a result of the sudden and rapid development of the local shark-fishing industry, which came about as a result of the demand for shark livers as a source of vitamins. In this type of fishing, gillnets are used at considerable depths, which require relatively large numbers of floats capable of withstanding the resulting high pressures.
>
> It was soon found that glass was the best solution to this problem as it would stand up better than cork or other available materials, so there was a sudden large demand for glass floats with no adequate source of supply to meet it. After unsuccessful attempts had been made to use beer bottles, which lacked sufficient buoyancy and suitable means of sealing, a local fishing supply firm prevailed upon us to take steps to supply this need.
>
> Since the days when we made these floats by hand, much progress has been made in the direction of high-speed automatic jars and bottle manufacture, so we decided to attempt the production of fish floats with modern auto-

> matic machinery and thus greatly increase the output and reduce the cost as compared to the hand process. As our glass machines will turn out floats at speeds as high as twenty-five and thirty per minute, we found that the problem of sealing them in a suitable manner at such a speed was quite a formidable one. However, after considerable experimental work, we succeeded in developing a very satisfactory automatic float-sealing machine and immediately went into the production of fish floats on a relatively large-scale basis.
>
> As far as we know, we were the first to make these floats with automatic machinery, and we do not know of any other company that has made any substantial number of them. We have so far made only the ball-shaped floats and in four sizes: 3½, 4½, 5, and 6 inches. We have produced them in both amber and flint glass, and you can identify them by our NW trademark stamped in the bottom. We have probably turned out in the neighborhood of two million of them so far.

During the period of the Second World War, Owens-Illinois Glass Company at Oakland, California, also used bottle machines to manufacture a quantity of the five-inch glass floats. There were two reasons for this: supplies from Japan, Russia, and Czechoslovakia were cut off by hostilities; and with the advent of bottom fishing, cork floats would not stand up under the water pressure. Besides, both Spanish and Portuguese cork was difficult to procure at that time.

The Owens-Illinois glass floats are also occasionally beachcombed. I found one in January 1962, at Sand Point Beach in Washington, south of Cape Alava beyond Lake Ozette. These floats are readily identified by the "Duraglas" in large raised script letters opposite the sealing button and the "I" in an "O" within a diamond trademark on the flat portion.

Glass floats are also reported to have been manufactured at this time by the Crystallite Products Corporation in Glendale, California; and during this same period the Pittsburgh Corning Corporation developed a six-inch glass float made

This is an experimental doughnut-shaped machine-made glass float developed by the Northwestern Glass Company of Seattle, Washington.

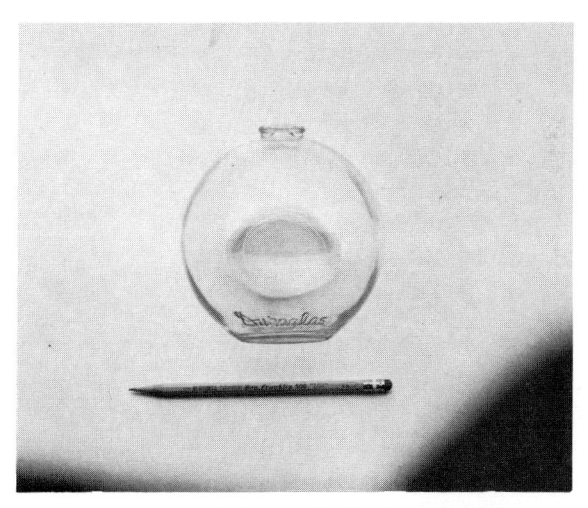

An American-made Owens-Illinois glass float showing similarity to bottle-shape with its flat bottom and sealed neck. These features are required for high-speed machine production. Note the trade name "Duraglas" near the bottom.

from two machine-pressed halves fused together. The glass used was their 1241K composition, formulated for excellent weather resistance. Although the float is probably the only one yet developed with a constant wall thickness affording superior strength at great water depths, it was about three times more expensive than the common bottle-blown type. Only a few hundred were made, and fewer were shipped for trial use, but occasionally they are beachcombed. We have two in our collection.

Perhaps the most unusual and most difficult-to-produce machine-made float in our collection is an experimental doughnut-shaped bottle-type float which Northwest Glass developed in 1949. Less than a dozen were manufactured. Unfortunately the market for glass floats deteriorated about that time, and mechanized methods for handling fishing nets required floats capable of resisting exceptionally high impact pressures; so this float was not put into production. It remains, however, as an outstanding production-process feat.

Pacific Coast beachcombers can easily get the impression that the Japanese invented the glass float. On the contrary, several nations used them much earlier in fishing Scandinavian waters and the cod banks of the North Atlantic. Norway began using them in the 1840s and Denmark in the early 1870s. Czechoslovakia and Scotland did not start until the 1890s. Japan, along with England, France, and Germany, began in the 1910s; Russia in the 1930s, and the United States in the 1940s.

Although not statistically confirmed, it has been my general observation that American machine-made glass floats are better sealed than the hand-blown Oriental type. If the number of floats found that are partially filled with water is any indication, then this impression must be true.

It is a rare thing to find an American machine-made float with a marginal seal allowing the water to enter; but a considerable number of Oriental floats have been found with some

Halves of the six-inch experimental Pittsburgh-Corning float were fused together from special weather-resistant glass but proved too costly to produce.

Here the fused halves of the Pittsburgh-Corning floats are shown with and without the attaching nets. These floats were used in the soupfin shark fishery of World War II and are now a collector's item.

water inside. Basically it is easier to seal a hole perhaps three-sixteenths inch in diameter for the American type, as against three-quarters inch in diameter for the Oriental type, particularly when the former is done with automatic machinery while the latter is individually hand pressed and rolled.

Hand-blown Japanese floats were known to be manufactured by the following companies in 1948:

Name	*Location*	*Monthly Capacity*
Sasa	Aomori	147,000
Sendai	Miyagi	62,000
Matsu Oa Oka	Yamaoata	72,000
Shimamura	Fukushima	50,000

They are now believed to be made also in Hakodate and Tokyo.

One of the major questions to be answered for the beachcomber is, "What is the future for beachcombers who seek the glass float?" In other words, approximately what reserve of floats is now in the Pacific Ocean for potential discovery on Northwest coasts? Using all the data at hand and what I consider realistic assumptions, I have concluded that the North Pacific Ocean still retains approximately ten million Japanese fishing floats.

Roughly, forty per cent of all Oriental glass fishing floats ever lost are probably still out there in the Pacific, continuously traveling the gigantic Kuroshio circuit, awaiting the chance storm that will drive them ashore on our West Coast. Part of this mammoth necklace of glass floats has been seen off Vancouver Island. Assuming this necklace to be thirty miles wide and all the floats spaced fairly equally apart, then each is spaced about one-quarter beachcombing-mile from another, which generally matches my findings of several floats per mile. Of course, this is an academic average and specific disturbing factors like the Oregon Maelstrom will improve local beachcombing results.

So, if you have yet to find an Oriental glass ball, just keep looking. Plenty of them are merely waiting to be thrust ashore, and plenty more are being turned out daily from the ovens of the Japanese glass blowers.

Ellis Nelso

The crew of the *Yury-Maru No. 8* relaxes on the afterdeck. In a typical sixty-day trip out of Samoa this ship brings back about seventy tons of fish. The catch averages around two tons per day from the 1,670 hooks of their line, which is put in and pulled out once each twenty-four hours.

CHAPTER 7

Japanese Fisheries and Fishermen

We were on the beach at Ocean Park, Washington. The sun was bright. Only a mild breeze from the ocean reminded us of yesterday's strong wind. We walked to one of the higher but still relatively low and flattened-out sand dunes to watch the sandpipers at the edge of the water. All around us was the inescapable roar of the surf. Down would crash a breaker to run slowly back along the flat, table-like beach, only to be followed by the next one, all sequenced according to some highly complex timetable which the sandpipers at least seemed to have figured out satisfactorily: they ran just inches ahead of the advancing wave and just inches behind the receding wave. The whole restful scene dispersed worries as quickly and effectively as the waves erase the sandpipers' footprints on the wet sand.

As I contemplated the broad, bright Pacific, I thought how, almost six thousand miles away, were the Japanese whose fishermen had unwittingly sent myriads of glass treasures to our Northwest shores and would likely send myriads more with their constantly expanding fishing operations.

Their major expansion in the Pacific waters during the 1920s was brought about by the advent of production gasoline and diesel engines. Conversion of their sail-powered fishing boats began with war-surplus engines. This increase in potential range of operations provided the Japanese with many new areas of the Pacific to exploit; and their joint-operations agreements with foreign countries made the exploitation practical. As export markets developed, larger vessels were built, and more fishing gear was manufactured.

Prior to World War II, almost all of Japan's overseas fishing and most of the home coastal fishing was accomplished by four large fishing companies. These organizations, with subsidiary companies, made Japan the greatest fishing country in the world and accounted for twenty-three per cent of the world's total volume of marine products. Over 1,000,000 fishermen and 350,000 fishing vessels were required to accomplish this. A brief look at these four fishing companies will reveal some enlightening details of their operations. They follow in order of size:

The Japan Marine Products Co., Ltd.
(Nippon Suisan Kabushiki Kaisha*)

The Nichiro Fishing Co., Ltd.
(Nichiro Gyogyo Kabushiki Kaisha)

The Ocean Fishing Co., Ltd.
(Taiyo Gyogyo Kabushiki Kaisha)

The Polar Whaling Co., Ltd.
(Kyokuyo Hogei Kabushiki Kaisha)

In 1928, Nippon Suisan KK, probably the largest in the world at the time, conducted 87 per cent of the total trawling, 99 per cent of the total crab processing in floating canneries, 40 per cent of the total deep-sea whaling, and 76 per cent of the total coastal whaling. This company had subsidiary companies in Argentina, Formosa, Borneo, the Philippines, Manchuria, and Korea; and it operated fishing vessels and floating canneries as far north as the Bering Sea and as far south as the Antarctic. It fished off Japan, in the East China Sea, the Yellow Sea, the South China Sea, the Bay of Bengal, off Australia, Argentina, and Central America. Although about one half of the fishing fleet was destroyed during World War II, Nippon Suisan today remains a leader in fishing products.

* Kabushiki Kaisha (KK) means "Company, Ltd."

Ellis Nelson

On the left is Captain F. Abe of the Japanese long-line tuna ship, *Yury-Maru No. 8*. On the right is Mr. K. Gunji, manager of the Japanese fishing fleet operating out of Pago Pago, American Samoa. The *Yury-Maru No. 8* grosses 112 tons.

Ellis Nelson

Bamboo marker poles and glass fishing floats are stored on the after cabin of these Nationalist Chinese and Korean tuna ships tied up in Pago Pago, American Samoa. The four kinds of tuna caught are Big Eye, Yellow Fin, Albacore, and Wahoo.

Nichiro Gyogyo KK was organized in 1914 and later became the sole company to operate off Kamchatka. Its operations included Karafuto, the Northern Kuriles, and Hokkaido, and its primary products were canned salmon and crab. It also had extensive boat-building yards, repair shops, machine shops, net factories, and other maintenance facilities. Before World War II, this immense organization included over 38,000 people, and 137 canning and processing factories. The company fleet then consisted of the following: 734 vessels, three of them (canneries) with 5,000 to 6,000 tonnage. In 1945, its activities were brought to a standstill by Soviet occupation of Karafuto and the Kuriles; however, the company has now again become a larger operator. Glass floats bearing its house flag or trademark occasionally show up for American and Canadian beachcombers.

Taiyo Gyogyo KK had extensive fishing operations in Japan, Korea, and Formosa in the 1930s. It also had a vast supporting organization of boat-building and repair yards, processing companies, canneries, cold-storage and ice-making plants, and box and net factories. As of 1946, this company was owned entirely by the Nakabe family and a few close relatives, and it is still a leader in fish products. Glass fishing floats containing the trademark of this company are rarely found and, when found, are a real collectors' item.

Kyokuyo Hogei KK had pre-war operations of domestic and antarctic whaling and some trap-net fishing; however, during the war most of the whaling fleet was destroyed. A year afterwards they constructed twelve whale-killer boats in the 100- to 200-ton class; but to my knowledge no glass floats bearing the company markings at that time have been beachcombed.

From the fisheries experiment station in the Hyogo Prefectural Office, I learned that, in 1960, there were 1,159 net-fishing boats in Japan. Of these, 603 belonged to large fishing companies, the remaining 556 being privately owned. Most of these boats were engaged in bonito and tuna fishing, and they operated their ocean fishery all year round. Their home ports were located almost entirely in Kanagawa and Shizuoka prefectures in the center of Honshu Island. In Kanagawa alone, there were 81 boats totaling 27,000 tonnage. Since most boat owners mark the floats with their own signs, there could be seven or eight hundred different markings on Oriental glass fishing-net floats.

The importance of fish in the diet of the Japanese people is well known; in fact most islanders have a predominantly fish diet. Statistics show that the Japanese average about four ounces of fish per day, whereas this amount is the average American consumption per week. It is thus easy to understand why the Japanese are expert fishermen, why they must employ a wide variety of fishing gear, and why this gear must be efficient and inexpensive. Yet only a few of the many different kinds of fishing gear require the use of glass floats.

All fishing gear can generally be divided into two main classes, nets and lines; and there are six types of Japanese nets:

Gillnets (to capture fish through entanglement)

Casting nets (to spread over a school of fish)

Lift nets (to lift or scoop from below)

Seines and trawls (to drag along the bottom)

Purse seines (to encircle a school of fish from the side and below)

Stationary or "set" nets and traps (to capture fish traveling along well-known routes)

Of the above types, the gillnet, trawl, and purse seine have used glass floats. The gillnet or "tangle-net" is held vertically by floats along the top with weights along the bottom. Usually gillnets are strung along in a straight line.

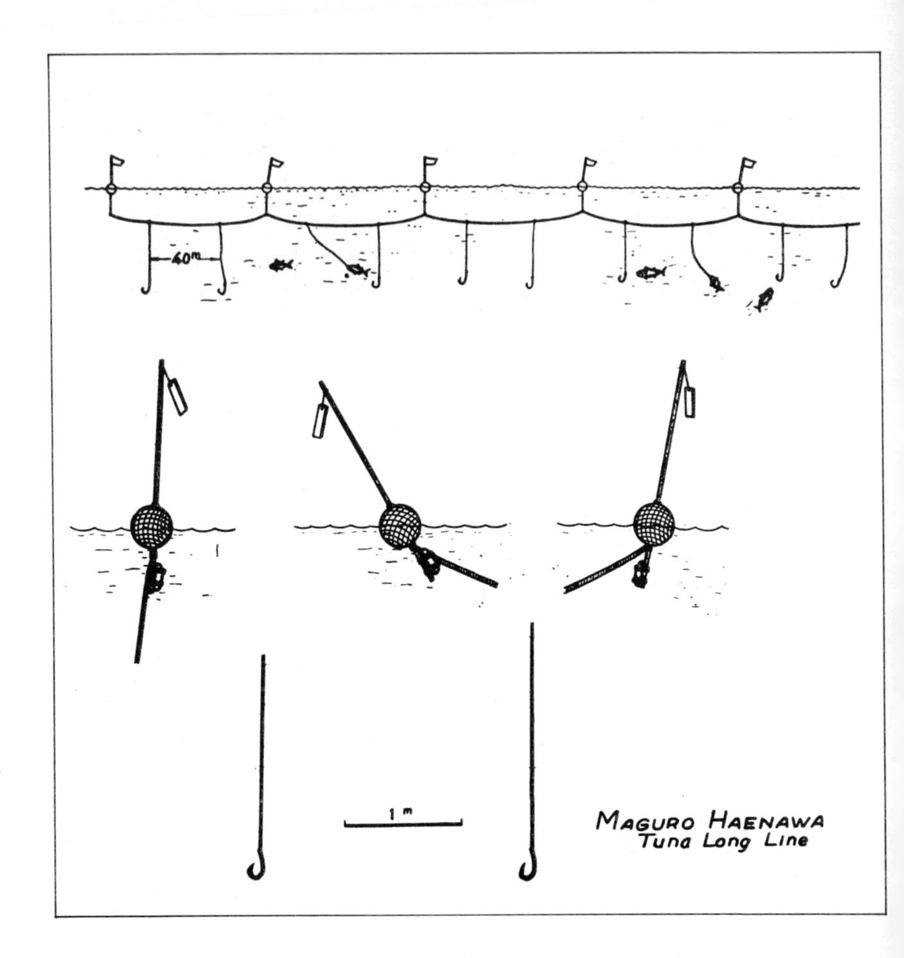

TUNA LONG LINE

Tuna long lines are held near the surface by glass floats, and a bamboo flag buoy marks the position of each float. Branch lines suspend the main or ground line, which carries the hooks. One end of the surface line is attached to the fishing boat and the gear drifts with the boat. Men in smaller boats watch the floats.

At Bristol Bay, Alaska, in 1961, the tangle-net operations of the Japanese were so extensive that one could cruise for thirty miles across an area that was marked every mile with buoys in both directions. This denoted a huge grid of underwater nets which completely covered the area in such a manner that the migrating king crab found themselves well fenced in.

Prior to 1947, Japanese trawls which used glass floats were the otter trawl and the small two-boat trawl.

The Japanese have developed to a high degree the long-line method for tuna fishing. This method involves stretching a single main line, often fifty miles in length, under the surface of the ocean to a depth determined by the length of attachment lines from glass buoys. Hundreds of fishing lines with baited hooks dangle from the main line. Each fishing line coiled separately for rapid handling consists of a swivel snap, a hemp line up to fifty yards long, and a wire leader containing the hook. The main line is made up of thousand-foot sections fastened together with swivel connections which minimize snarling. Attachment or buoy lines, which connect glass floats to the main ground line, ride the long surface line. To the nets of the floats are attached bamboo poles carrying signal flags.

It is reported that occasionally a large tuna will dive after being hooked and pull the nearest glass float so far below the surface that it is crushed by water pressure.

"Voyage of the Lucky Dragon"

I once wrote to a Japanese fisheries representative in Tokyo to ask for references covering a typical description of Japanese fishing and was told that one of the best descriptions was in our own Dr. Ralph Lapp's book, *The Voyage of the Lucky Dragon*, Harper & Row, Publishers. Following is his authentic account in Chapter I of the laying of a long line by Japanese fishermen near the Marshall Islands:

Kanagawa Prefecture Fishery Co-

Aboard a Japanese tuna clipper a fisherman ties a bamboo pole to a long-line marker float. Note additional poles in background.

K.P.

Tuna long-line marker poles and floats about to be attached to the long line. Note the phonograph records.

Electric lights played over the ship's stern as the fishermen started throwing the long line . . . baited hooks two hundred or more feet below the ocean's surface in a long, connected line suspended from hundreds of floats. As the *Lucky Dragon* headed on a northwest course, the crewmen took the No. 1 buoy, a twelve-inch-diameter metal sphere topped with a battery-operated lamp bulb, and dropped it over the stern. Buoy No. 1 was coupled to a float line, the end of which formed the beginning of the main or long line.

The glowing lamp bobbed away in the choppy waves as the boat proceeded on its course, and the fishermen kept watch on the main line as it unwound from a rectangular wooden box placed on the fantail. As the line snaked out of its container, crewmen kept bringing boxes of line and bait fish to the stern. One crewman baited the strong steel hooks, skewering eight-inch frozen fish through the eyes. The bait fish consisted of samma (mackerel-pike) which abound in the coastal waters of Japan.

Five branch lines were attached to one section of the main line, usually at intervals of about thirty yards. For every three hundred yards of the main line, a float line was clipped on and a green glass buoy thrown overboard on the end. Two buoys no larger than basketballs suspended a "set" of five baited lines, which settled slowly in the water, taking an hour or so to reach their final depth. The hooks nearest the buoys remained at a more shallow depth because of the sagging of the main line at its middle point.

Having thrown over twenty sets, or one hundred baited hooks, the fishermen substituted a lighted metal float for the ordinary green glass buoy. They reckoned their catch in terms of fish caught per hundred hooks or per "basket," using an old term which originated when a basket contained all the lines for one hundred hooks. The floats in between the lighted buoys carried a six-foot bamboo pole topped with a hemp palm or a square foot of red and white cloth.

Laying the long line took quick action and co-ordination

K.P.F.C.

Boxes of frozen bait used on long-line tuna hooks.

K.P.F.C.

Japanese fishermen attach baited hook-lines to long-line, as it is being payed out over the stern.

K.P

Working the long line on the star side. Note lights for night work

on the part of the fishermen, for the *Lucky Dragon* maintained its seven-knot speed as the line was paid out over the stern. Several times the line would go taut as the men failed to keep up with the ship's speed. There was much cursing at such times, but for the most part the men worked in silence, keeping their eyes on the job at hand.

It took almost four hours to throw the lines, and the men knocked off to have breakfast while the boat floated with its engine stopped. The morning sun revealed a long avenue of dancing buoys, stretching away from the boat until they were lost in the waves. Actually the line stretched beyond the visible horizon for a distance of about thirty miles. Beneath the waves over 1,500 baited hooks dangled, waiting to tempt some roving denison of the deep.

The crewmen worked in shifts, taking time out for mugs of steaming hot tea of the type grown in Shizuoka near their home port. For breakfast, the men had bean soup and rice, washed down with the bouillon-like tea. Some of the crewmen ate on deck, using chopsticks to "eat" their soup, selecting the solid pieces with the chopsticks and slurping down the liquid from the plastic bowls. They speculated about the first fishing of the year and how it would be a good sign if the catch was good. . . .

In the previous season the fishing had been excellent, and the catch, mostly tuna, had brought about $15,000. This time though, the thirteen hours' labor of hauling in the long line netted only fifteen fish. The crew grumbled and blamed the fishing master for the location he had picked, but there was limited fuel and they were approaching their maximum radius. Furthermore the weather was getting worse. Once more they threw the lines. Then they were plagued by engine trouble. Upon pulling in the main line they found it had been broken, perhaps fouled by a coral outcropping. Somewhere forty miles of fishing gear lay on the ocean floor—a tragedy for the fisherman but a virtual treasure for the beachcomber.

British Columbia Government Pho

Looking northwest toward Long Beach and Wickaninnish Bay on the west coast of Vancouver Island. To the upper left is Tofino airport and beyond would be the village of Tofino. Along this coastline is some of the finest beachcombing of western Pacific Ocean shores; and it was near here that Captain Robert Gray and his crew of the ship *Columbia Rediviva* spent the winter of 1791-1792, before going south to discover the Columbia River.

CHAPTER 8

Tofino Interlude

It was black as night when I left for Tofino on one early August morning. A few patches of fog in the lower areas reminded me that fall was just around the corner. A sliver of moon to the east suggested that the salmon fishing should be good. After an hour the northeastern sky started to brighten, and I noted the triangular silhouette of Mt. Pilchuck as I drove along the broad, flat area north of Snohomish. I was scheduled to board an airplane for a beachcombing vacation where I knew I could find glass floats.

I parked the car at the airport, checked in my baggage, and a short time later the small amphibian took off. The pilot had said the coast weather was not good. The green tide-flats west of Vancouver airport had turned to dark blue. As we climbed, I was glad I had worn my woolen jacket. Soon we came to Vancouver Island; below were Cowichan Bay and Nanaimo—looking asleep in the morning haze.

A few minutes later we passed Port Alberni. Surrounded by mountains, the huge mill was belching its usually heavy, white smoke. Lake Nahmint, the steelhead fisherman's paradise, came into sight, but by then the coastal fog had poured in from the ocean. The only other passengers—two loggers who were returning after a night on the town — were sound asleep.

As we flew on, I knew we would not land at Tofino but at Tahsis, an island lumber operation another eighty miles be-

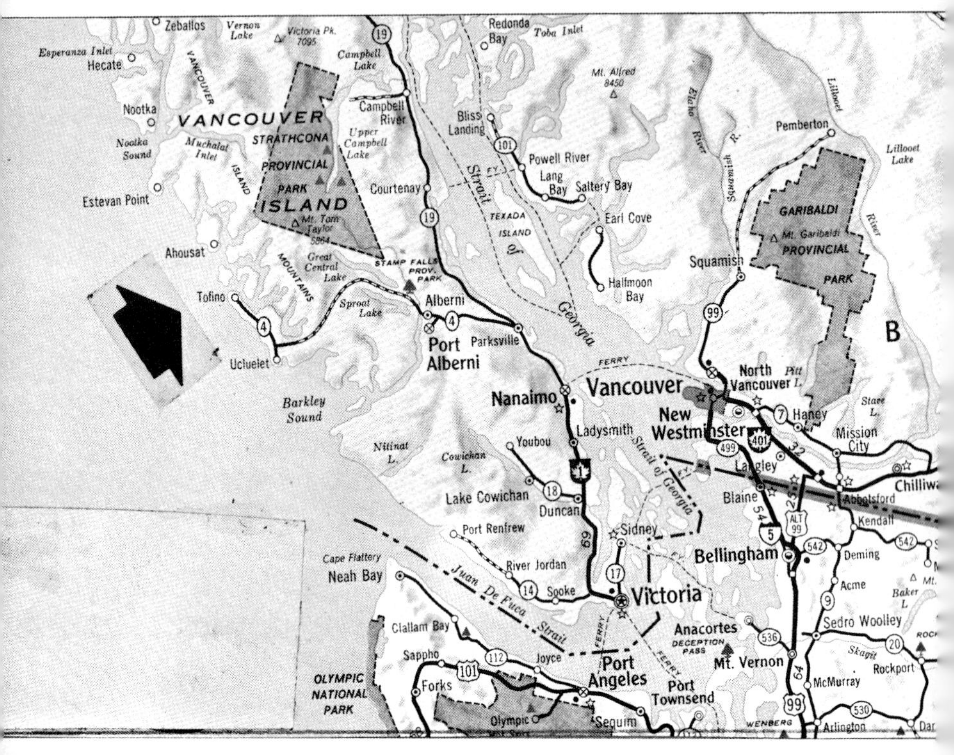

Lower Vancouver Island, showing the Tofino-Ucluelet area. Midway between them is Long Beach, a magnificent stretch of hard white sand. North of Tofino is low, wooded Vargas Island, and beyond Vargas is Flores Island, where the Indian village of Ahousat is located—all happy hunting grounds for the treasure seeker.

yond. Never before had I flown so close to mountain peaks. Soon we threaded our way down a valley, just comfortably clearing the treetops for the final approach over the mill and a smooth landing in the inlet. The pilot said that we would try for Tofino after a two-hour weather hold. Already my vacation leisure had started, for here in this small lumber town I was forced into doing nothing. I could accept the weather delay cheerfully.

Later we took off from choppy water and flew down the channel. First came Nootka Island with Friendly Cove (Nootka Sound) barely off the wing tip. Many times I had read of this historic place where, in the late 1700s, British, Spanish, and Yankee fur traders had brought their ships. Yet how peaceful it looked now, giving no clue to the greed and hostility rampant here in the years of exploration. Down there living in one of the small houses was Ambrose Maquinna, a direct descendant of powerful Chief Maquinna, who captured the good ship *Boston* and massacred all the crew but John Jewitt, the armorer; and Thompson, the sailmaker.

While these snatches of history were flashing through my mind, I suddenly became aware that we were passing over a series of beaches offering prime beachcombing; I had often studied the charts of these particular areas. Because of their inaccessibility, I had even investigated chartering a helicopter to go in there. Now here were these beaches right under the wing tip. Estevan Point also slowly went by. I had once written the lighthouse keeper there about his beachcombing of glass floats and he had replied it was good.

Next we flew over Flores Island, and Hobbs, Burgess, and Vargas Islands, all familiar haunts of mine these past years. After Lennard Island Lighthouse, we flew right over Cox Bay. I looked down at the shore and spotted Don and Ruth Close's beachplace—my destination.

Forty minutes after landing at Tofino, son David Close was skippering me by outboard boat toward Vargas Island and its

British Columbia Government Photo

Hiking along Brady's Beach near Bamfield, Vancouver Island.

Here is the way the Don Close family spent part of a leisurely sunny August day, making furniture for their campfire site on Cox Bay.

outer beaches. We anchored in a secluded and protected cove, then headed west over the rock outcroppings, switching back into the thick underbrush of salal and Oregon grape at the deeper chasms which couldn't be crossed. An eagle screamed at us because we were intruding upon its private domain; atop a spruce we saw its huge nest. Beyond, the long, wide beach came into view.

Six hours and five miles later—thanks to David's uncanny eye—we had fourteen Japanese glass floats, two Japanese saki bottles (now in demand by amateur ceramists), five fishing plugs, two heavy-duty fishing lines, a Japanese light bulb, and a colorful cedar crabpot float. We returned quite late but Ruth Mary Close had a hearty outdoor supper waiting: baked salmon, lettuce salad, baked potatoes, cherry pie, and tea. We lingered around the huge open fire and talked. When Don added a log, the sparks rose high into the still, black night.

Later, as I was bedded down in my sleeping bag under a scrub spruce, I reflected on the events which had started some eighteen hours earlier. The first day of my August vacation closed with the light from the Lennard Island Lighthouse flashing through the trees, and I fell asleep knowing that Vargas Island would often be an important part of my future beachcombing travels.

Delano Photographic

A portion of the Pacific Ocean coast near Depoe Bay, Oregon. Note the protected anchorage for small craft. In the foreground are rockbound shores; however, beyond in the upper part of the picture, are the flat sand beaches stretching past Gleneden Beach, Taft, Nelscott, Delake, and Oceanlake, where the beachcombing is excellent.

CHAPTER 9

Where To Find Glass Floats

Glass floats, in theory, should wash up on the shores of the entire North Pacific Ocean Rim. Actually they occur where major currents are forced into sharp turns; floats and other debris are then more easily caught in boundary shore currents and driven inland by local storms. Such turns are located off Dutch Harbor in Southwest Alaska, the Queen Charlotte Islands, Vancouver Island, Oregon, Acapulco in Mexico, and Luzon in the Philippines.

Seasonal movements of main ocean currents also help explain why, at times, a particular beach is favored with debris including the glass balls. The great Black Stream, Kuroshio, changes its latitude continually, according to the season, and utilizes the local storms to carry its fruit onto flat beaches. For several years I have observed Vancouver Island's west coast, both before and after storms, with considerable assurance that this must be true.

Where some of Kuroshio's current fans out into the North Pacific Drift, this seasonal movement has been scientifically plotted in a study by the Pacific Oceanographic Group at Nanaimo, British Columbia. Their report, dated May 1958, describes the release of twenty-five thousand sealed drift bottles from weather ships approximately a thousand miles west of Vancouver Island. The releases covered a full year. The location and dates of bottle retrieval give directions and velocity of travel. This report has direct correlation to probable paths of glass fishing floats in this region. It is particularly significant that the effect of the wind on bottle drift is considered negligible. The report states in part:

> It is recognized that the bottles, floating on their sides, cannot indicate any more than the drift at the sea surface. It has been suggested that these bottles, without drogues (anchors), are influenced to a considerable degree by the wind, independent of the movement of the water, but it is reasoned that this is improbable. The sea surface is normally rough (waves one to three feet high).
>
> It has been observed that the bottles are usually smothered in foam on the wave crests where they would be exposed to the wind. They are only exposed in the smooth trough of the waves where they are protected from the wind. Hence it is suggested they advance with the surface of the sea waves, rather than the wind.

And so it is with our Oriental vagabonds. No wonder they collect many forms of sea life during their trip.

On what specific beaches of the Pacific Coast can glass floats be found? Since the main ocean currents that bring them change their locations seasonally and annually, any flat, sandy beaches from Acapulco, north past Vancouver Island and the Queen Charlottes are potentially good places. One season may be better at Newport, Oregon, than at Cape Alava, Washington, one of the richest areas. Another season may favor the Alaskan Coast. Now to describe some of these beachcomber haunts. . . .

Washington

Almost all of Washington's outer beaches produce glass floats of many sizes and varieties. From Tatoosh south to Cape Disappointment at the mouth of the Columbia River, there is a coastline distance of 150 miles, consisting of 130 miles of beaches and 20 miles of steep, rocky bluffs. Perhaps the Long Beach area between Cape Disappointment and Leadbetter

Point is the best known because of its ready access and high float yield. This may be due to the Oregon Maelstrom and the spoiler action of the Columbia River jetties, which extend several miles into the open sea.

Cape Alava, just south of Cape Flattery, reaches far out into the ocean to catch its beachcombers' treasures: it is the most westerly point of the United States mainland, excluding Alaska. A neighbor told of a successful winter outing there. After hiking in on a February night and making camp in an abandoned shelter used by occasional Indian fishermen, he woke up the following morning to six inches of snow. He and his friends hiked north along rocky beaches, where rocks of all sizes extended far out into the surf.

Here where there was very little driftwood or sand, it was difficult to imagine anything not being smashed or broken by the sea. Yet, at an extremely northern cove, all in a bunch, they found more glass floats than they could easily carry out, some the size of basketballs. It takes only two twelve-inch floats—in a pack already containing twenty pounds of sleeping bag, food, clothing, and some water—to add another eighteen pounds. That forty-pound pack gets rather heavy toward the end of the eight-mile trek back to where the car is parked.

Although it is well known how floats will travel several thousand miles across a large ocean and be tossed up onto adjoining beaches, it is less widely known that they will also travel into inlets, carried there by currents and tides, hundreds of miles from the ocean. In 1937, a six-inch float was found on the beach at Point Defiance, near Tacoma, Washington, which is about a hundred and sixty miles inland from the Pacific Ocean proper. And three were discovered tied together on the beach near Sachet Head on Whidbey Island in 1960. Glass floats have also beached at Hood Canal, a navigable inlet in Puget Sound within the Strait of Juan de Fuca; and at the fiords of Norway and Alaska. These latter narrow inlets of the sea, with their high, rocky banks, are very successful traps.

During an annual clam hike in 1960, we found a float near South Point in Hood Canal and another at Port Gamble, also in Puget Sound. In fact, chances of finding floats along the shores of Puget Sound might be better than out on the Pacific beaches; most people at the beaches have their detection senses tuned up, while relatively few people are on the lookout in inlet areas. Another reason is the special guardian angel that Puget Sound beachcombers have.

This "angel"—who will remain nameless—happens to be a Seattle-based fisherman. Upon entering Puget Sound after each Alaska trip, he drops overboard a dozen floats that he finds up in Bristol Bay. He told me that he once had a neighbor who tried to beachcomb a glass float but, in his many attempts, was never successful. One day when this fisherman was harvesting hundreds of salmon gillnet glass floats from a particularly good beach, he thought of his neighbor who had yet to find one. So the idea was born to re-distribute some of this wealth. The neighbor has since moved elsewhere, but our fisherman friend still engages in this benevolent gesture.

A proprietress of an Ocean Park, Washington, restaurant tells that after a hard day over the stove in July, she and her husband decided to go out on the beach to relax. Immediately upon turning off the beach approach, they found five small floats camouflaged in the brown foam. (This brown foam is produced from the surf's pounding the sand into a fine, brown powder. Some of this foam breaks away and is blown along on the sand, picking up an outer layer of sand which makes it similar to a potato in size and color—a good camouflage.) A few minutes later, the husband found a large long-line tuna float, which now adorns their shop. It has the Japanese symbol for "green" on it.

The experience was unusual in four ways: first, July is usually not a good month for beachcombers; second, it is the

Hjalmar Brenna

Hours after reported beaching of dimension lumber near Ocean Park, Washington, in January 1966, beachcombers arrived in cars, pickups, jeeps, and trucks. Many braved the surf to salvage the longer planks.

Wayne R. O'Neil

Cecil Holman with a morning's haul of large glass floats found at Long Beach, Washington.

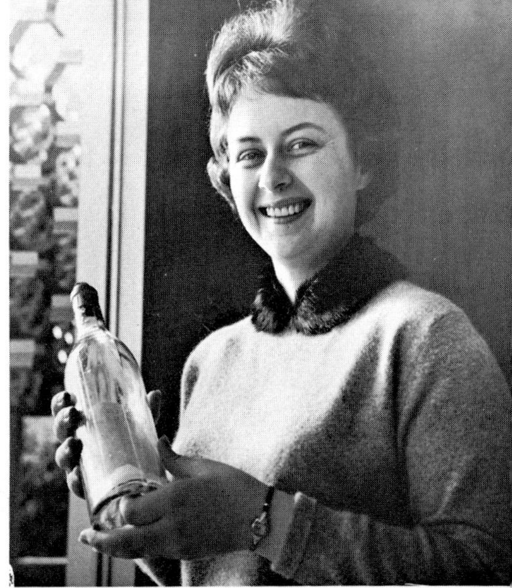

Boeing News

Comely Boeing stenographer found a sailor's note in a bottle while she was beachcombing Washington's Lake Ozette Beach.

height of the tourist season; third, I know of only two floats ever found with that particular marking and have no way of knowing what the symbol stands for; fourth, the discovery was in early evening after the vacationing beachcombers had had ample chance to pick the beach clean.

Grayland, Washington, has been the site of an annual beachcombers' derby which is widely advertised west of the Cascades. Besides awards for driftwood displays, prizes are also given for the largest Japanese glass fishing float found by a family group.

Beachcombing within Willapa Bay and Grays Harbor apparently is not nearly as good as on the outer flat beaches. The Westport and Grayland beaches are excellent. In the January 1966 beachcombers' bonanza of large glass floats, several findings of around a hundred floats were reported from the Westport area. North of Grays Harbor is a long sand beach stretching almost undisturbed for more than twenty miles from the Ocean Shores project north past Ocean City, Copalis, Pacific Beach, and Moclips to the Quinault Indian Reservation at Taholah. All of these areas are well known for their glass-float harvesting.

During the 1930s, one could ride the bus from Seattle out to Copalis Beach for a modest sum. It was here in this general area during the Depression that many people built small shacks from beachcombed lumber and subsisted principally from the beach. Many a glass float found by these "Depression Beachcombers" was sold or traded for groceries.

Oregon

With its nearly five hundred miles of coastline—including close to three hundred Pacific Ocean beach miles—Oregon has been considered the center of glass-ball beachcombing. Probably more glass floats are harvested from this coastline than from any other part of the Pacific Ocean Rim, primarily because of the large population nearby. Also, the north and south sea-

Brenna Photo

When the lumber barge *George Olson* was wrecked on the Columbia Bar, January 30, 1964, much of her 3.5 million feet of lumber was beachcombed off Cape Disappointment.

Burford Wilkerson Photo

The violent wind storms of January 1966 washed these objects ashore near Tillamook at Netarts, Oregon: floats, two bottles, a huge Japanese light bulb, and a Norwegian rubber buoy. Finders (from left): Mr. and Mrs. Kent Drake, Warder Meyers, and David Coffman.

sonal movement of the Japanese Current sweeps past much of the Oregon Coast. Because Oregon is such a lengthy beachcombers' paradise, I will consider it in three parts. The northern section—starting at Astoria on the Columbia River and extending fifty miles south to Tillamook on the southern end of Tillamook Bay—has many excellent places for beachcombing. For example, it has been estimated that the sand spit at Nehalem generates a yearly average yield as high as fifteen floats per mile per month. The "float find" in this region is reported to be similar to that just north of the Columbia. Likely both areas are subject to the Oregon Maelstrom, that slowly rotating vortex alleged to be located off the Oregon Coast.

The Nehalem beach is the site of an early Spanish shipwreck which some historians believe to have been the *San Francisco Xavier*, wrecked here in 1705—from which large quantities of beeswax washed ashore. Recent tests of chunks of this beeswax, however, indicate a date some two hundred years earlier. History records that the local Indians traded pieces of this beeswax with the first Whites. Such a vessel, if rendered helpless off the Philippines by a typhoon, would drift in the Kuroshio Current north past Japan and then eastward across the Pacific Ocean, directly into the Oregon Maelstrom and finally drive onto Nehalem Spit. Numerous Oriental junks are known to have washed up on these same Oregon beaches. The much-sought-after Neahkahnie treasure, believed to be buried in the Nehalem vicinity, was reported to have been carried ashore from another early sailing ship wrecked here. Few beaches in the world can match Nehalem Spit for mystery, romance, and adventure; and, largely responsible, is the Japanese Current, Kuroshio.

One of the regular beachcombers of Nehalem Beach is Mrs. Robert Wise, who walks the seven-mile sandy stretch almost every day. Most of the glass floats she picks up she sells in her gift shop located in Manzanita, Oregon. Her most

Burford Wilkerson

Beachcombing near Tillamook, Oregon, provides a wide variety of items—all just for the looking. Here is a typical find of interesting things from at least three countries. From the left: Mold-made floats, stave with carved oriental letters, cans of emergency drinking water, cheese box with French markings, wooden shoe, wine bottle, driftwood, and wood with Japanese marks.

Verna Slane

This picture of Boiler Bay, Oregon, taken at low tide, reveals the old boiler of the wrecked ship, *J. Marhoffer*, which gave the bay its name. Parts of the ship—lost May 18, 1910—continue to be exposed to beachcombers by wave and tidal action. The entire picture area will be under water in a few hours as the tide comes in.

memorable beach hike occurred on a bright moonlight night when float after float came in through the surf to roll off the last wave. As each float came to rest on the flat sand, it shone in the moonlight like an individual giant pearl lost from a necklace—the Kuroshio necklace.

In March 1966, we beachcombed several of the beaches between Newport and Tillamook, Oregon. After dinner at the Salishan Lodge, we studied their maps and decided to try the local area the following morning from Gleneden Beach north along the sand spit. Dawn found us walking the high-tide line in a hailstorm driven by a nasty southwest wind. The previous tide had deposited Velella on the sand in a long, scalloped border, so the searching area was readily defined. Within a few minutes, among the Velella I found three small glass floats which were covered with marine growth. As it grew lighter and the hail lessened, the beach suddenly seemed to come alive with people. I knew the Oregon beaches were well combed and here I saw ample proof. Beachcombers that morning picked up some ten small-sized floats on the two-mile stretch. I had found three of these in front of the Salishan Beach Longhouse.

Around noon that same day, just after the second high tide, I worked the beach south from Pacific City, Oregon, which is about another twenty-five miles north. While walking around the driftwood upwind into that same cold rain and hail, I noticed an intrepid soul approaching me from the south. After our greeting he said he lived nearby and had been out since dawn. He produced a single small float from under his waterproof jacket. I complimented him on his find and after a short conversation we parted. I continued south for another half mile, but soon lost heart knowing that his pickings had been slim. The cold wind seemed not about to weaken, so I turned downwind, took a few photographs, and made a beeline for the warm car and Elaine's lunch.

L. F. Anderson

Looking northeast of Gleneden Beach, Oregon, where beachcombing is good all year around. From left to right: Siletz Bay, Salishan properties, and Gleneden Beach. The main road is Highway 101.

Burford Wilkerson

Evidence that there is an Orient across the Pacific Ocean from Tillamook, Oregon—the cocoanut at the left has a good growth of gooseneck barnacles. The 100-watt Japanese light bulb is almost as indestructible at sea as the glass fishing float shown at the right. The smaller barnacle growth in the glass float indicates it had a shorter period at sea.

The central region of Oregon, from Tillamook to Coos Bay—including Oretown, Newport, and Winchester Bay—has long been known as fine beachcombing country. There are many excellent float collections in this area, and I am told that some people here, over the years, have stored floats in attics or basements with the thought that in time they would become more valuable.

Here on Oregon beaches, as on many other coastal beaches, the currents may deposit glass floats in a local region, say as short as one hundred yards. Then the beachcombing may be poor for several miles. Often there will be no visible change in the beach line itself, but the beachcomber who continually works a long stretch of beach will find, through experience, the better spots.

Oregon's southern coastal region, from Coos Bay to Brookings, also produces well for the beachcomber. The big coastal blow on Columbus Day of 1962 reached a hundred miles per hour at Cape Blanco, according to reports, with excellent beachcombing following. My family and I have beachcombed virtually all of the Oregon beaches and this is one of our favorite areas.

During our 1964 Thanksgiving-vacation beachcombing trip, we visited Cape Blanco, the most westerly point on the Oregon Coast. We took our bearings from the lofty and strategic position of the lighthouse and found that the beaches running south appeared almost inaccessible because of the steep and slippery three-hundred-foot clay banks, whereas northward these banks flattened out and the beaches were readily accessible.

Our beachcombing near Bandon, twenty miles north of Cape Blanco, provided an identical list of things we had found on Vancouver Island: hatch covers, parts of broken boats, stumps, logs, planks, and plenty of kelp. Beachcombing in the region of Port Orford is reported to be so well organized that airplanes have been used to locate where the floats are arriving; the pilot then by radio directs his partner in an automobile to the best areas.

A retired Navy Petty Officer from Coos Bay wrote:

> Beachcombing here for floats is best in late fall to early spring, as that is when we have the heavy surf. We believe that about the same number of floats come in every year, depending on the wind and surf.
>
> I have given up beachcombing for floats as the sand buggies and four-wheel drive vehicles are too much competition. About all I beachcomb now is lumber, also wood for the shop and for the Franklin fireplace in the front room.

It is my conservative estimate that, since 1925, more than five hundred thousand glass floats have been beachcombed from the Oregon Pacific Coast beaches. And, to add to the pleasures of beachcombing there — the Oregon coastline is among the most beautiful in the world.

California

California has one of the longest coastlines bordering the Pacific Ocean, but it also has a population of some fifteen million, and the beaches near the great population centers are much frequented. Here, beachcombing as such is more a matter of mere luck. The rule still holds, though: the beachcomber should seek the remoter beaches so that he will have a better chance of being "first" to comb. Because the southbound California Current—which moves parallel with California's northern coast—begins its southwest curve out to sea in the region below Monterey, there are more glass-float beachings in northern California than in southern California.

A story is told by a friend about finding one in San Francisco Bay:

> In the early 1940s, the western side of Alameda Island was lined with an unbroken row of waterfront mansions.

The shoreline was inaccessible except for a rare break where an occasional street dead-ended at the beach. During the winter of 1941-1942, I was attending school at the Oakland Airport and living in a boardinghouse—one of many in the city of Alameda. One Sunday, my roommate and I drove from our boardinghouse down to the waterfront at the end of one of these dead-end streets. We were sitting on the sea wall, looking at the view and watching the waves, when we noticed two small boys some distance away playing with what appeared to be a bluish-purple balloon which had been washed ashore. They were starting to throw rocks at it when I realized what it was.

I ran down the rocky beach toward them, calling to them to stop. The object turned out to be a beautifully colored Japanese glass fishing float, about 14 to 16 inches in diameter. Having been raised in Seattle, I had seen many of the glass floats there, as well as in various curio shops up and down the Pacific Coast. Besides, I knew my mother would be very glad to have this one. I offered the boys thirty cents for it and they seemed happy to make the trade. As I recall now, I took it to Seattle on the train during the Christmas holidays and presented it to my mother as a Christmas gift. I do not believe there were any markings or characters on it; however, that was quite a while ago, and I don't recall if I even really looked for any.

The most remarkable feature of this find lies in the path the glass ball must have followed through the narrow and treacherous waters of the Golden Gate, past Alcatraz and Yerba Buena Islands, and finally ashore at Alameda.

Vancouver Island

The wind-driven waters of the Pacific have traveled eight thousand miles when they reach outer Vancouver Island, and the surf that batters this stretch of open coast is reported to be the worst of any beach located north of the equator. The island is about 275 miles long from Victoria to Cape Scott, but there are approximately 336 west-shore coastal miles because of the irregular coastline. Of the total shore miles, some ninety contain flat, sandy beaches; and the best stretch of these, some thirty-eight miles, is within a relatively small area from Estevan Point to Ucluelet. I have covered most of this area on foot and can attest to good beachcombing. A full gamut of different and interesting Oriental debris can be found here because of the long, open fetch, the ocean currents, and the relatively high incidence of stormy weather. Such things as Japanese boards, Chinese crates, and Polynesian mats are not uncommon.

The remaining fifty-two miles of sandy beaches have difficult if not impossible access; it follows that exceptionally good beachcombing can be found there. Occasionally these areas are visited by helicopter. Here is the experience of Del Dinger, of Long Beach, Washington, regarding such an inaccessible beach:

> In 1942-43, I was lucky to secure a job with the United States Coast and Geodetic Survey. I was then a Seaman First Class, formerly of Tacoma, Washington. Our job of surveying found us in very out-of-the-way places, some only accessible by small boat. One such place on the west side of Vancouver Island, on a very private beach close to the Carmanah Lighthouse, was a beachcomber's haven.
>
> I was put ashore there to stay for three days, but owing to very stormy weather my stay was extended to seven days. A Greek ship had blown up shortly before this off Tatoosh Island (the *Coast Trader*, torpedoed by a Japanese submarine, January 7, 1942), scattering debris all over the west side of Vancouver Island, including such items as a desk, chairs, baby grand piano, etc.
>
> In my stay of seven days, I also picked up over 600 glass balls, brought in by the same storm, and had them piled in little piles all over the beach at the high-tide mark. I found red, purple, amber, green, and odd-shaped ones, such as the rolling-pin type. When my boat came to pick me up off the beach, I could take just a few floats with me. I took the odd-shaped and different-colored ones only. It broke my heart to leave all the rest behind.

B. M. Simpson, of Port Alberni—a pilot of British Columbia Airlines—tells a story of caches of floats along the beach between Clo-oose and Port Renfrew, Vancouver Island: A telephone maintenance man had the job of walking the line by the ocean some twenty-odd miles, to check the telephone line. In his walks he regularly collected glass floats as they arrived on high tides. Often he couldn't carry them all, so he left piles of them at the base of the telephone poles; not at every pole but at many along the way. Unfortunately, he died before gathering them. I asked how to get into the area and was advised to drive to Nitinat Lake over the McMillan, Bloedel logging road, go by boat across the lake, and then hike several more miles through the brush to the beach. The floats are alleged to be stacked against poles within the first ten miles south from Clo-oose. . . . Del Dinger later added to this story:

> When I was on that beach, I met that telephone man on his route one day. I had quite a talk with him. He had a rough and lonely job. He carried a pack with tools, food, and a gun for protection from the many animals. It was a really wild and deserted country. He walked on the beach where he could, but some of his trail was boardwalk over the marshes and swamp. He crossed the rivers by cable in a cart on wheels, hand over hand. He also had a shack every so often to stay in over night, and he always left some food for his return trip. As I remember, the distance he walked on this telephone line was closer to fifty miles than to twenty.

Simpson also told of the legend that an old ship's cannon exists in an abandoned Indian camp on Kennedy Lake. If so, this could be one from the ill-fated *Tonquin*, which sank at Clayoquot Sound in 1811. However, it seems incredible that the cannon could have been towed against the normal flow of the Kennedy River. Certainly no small rowboat could make way against that current. In the spring of 1960, I tried rowing a dinghy into the inlet past the abandoned salmon cannery and held my own for a few minutes, but finally had to give in

Beachcombing the easy way by—helicopter—on the outside of Flores Island, B. C.

to the strong current. A large Indian canoe, the type the Clayoquots and Nootkas used for their whaling, might have made it. Whether the Indians could haul a fifteen-hundred-pound cannon is questionable, even if they used more than one tug canoe.

This legend of the *Tonquin* cannon could also be based on one of the wooden dummy guns that was aboard; the ship carried ten cannon plus a number of sham guns to impress strangers. A wooden cannon could have been taken in over a trail to the lake.

Queen Charlotte Islands

The Queen Charlotte Islands are the home of the Haida Indian nation. Because of the large red cedar that grows there, the Haidas are believed to be the originators of the totem pole, the well-known symbol of Indian villages from Seattle, Washington, up to Sitka, Alaska. The Charlottes are as far north as Prince Rupert, British Columbia, and are separated from the main Canadian Coast by some fifty miles of water.

The comparative isolation of this part of British Columbia, plus my study of detailed charts of its beaches, prompted Elaine and me to take an exploratory beachcombing vacation there late in October of 1963. The possibility that these islands might be a beachcomber's haven had been tucked away in the back of my mind for a long, long time.

We left our Mercer Island home by car in the early morning in pitch-black darkness and a light rain. Three hours later we arrived at the Vancouver, B. C., airport in the midst of a small windstorm. An airline employee said it had been blowing hard all night, and prospects for a smooth flight looked slim. We boarded the plane while it was still dark and soon were flying in the overcast. We never did see any of Vancouver Island. Elaine took a book from her huge handbag and started to read me some John Steinbeck. I think she was trying to take my mind off the *Fasten Seat Belt* sign which was continually reminding me of the rough air we were in.

After a bumpy breakfast of French toast, rolls, and coffee, I peered at the gray nothing off the wing tip, wondering just how far out to sea we were. Suddenly, without a waver, we flew out into the clear. The sun was tipping the tops of clouds and we could see the ocean through a lower layer of scud. I hurriedly looked back and found we had flown out the side of this storm, for there was the wall of rain extending to the water. Immediately below were whitecaps which looked quite disturbing, even from our vantage point at twenty thousand feet. Perhaps this early fall storm would bring in some of the glass floats that hadn't been beached during the past mild year.

We landed at Sandspit airstrip under a dark ceiling, and the small amphibian airplane which was to meet us was not there; the interior weather had forced the pilot to cancel our rendezvous. Here we were on Moresby Island, still a hundred miles short of our destination, the most northerly village of Masset on another of the Queen Charlotte Islands. Thanks to a bus, water taxi, land taxi, and a chance meeting with a home-bound Masset resident, we were providentially deposited in Masset—twelve hours from our home on Mercer Island.

Our friends at the small Haida fishing village of Masset on Graham Island were Phil and Norah Burton. Phil, a pensioned British World War I pilot, and Norah, his delightful Yorkshire wife, proved fine and companionable hosts. Their relatively uncomplicated living captured us immediately.

On the morning following our arrival, it was decided to beachcomb an area that Norah had known since she was a young girl. We drove past the site where her architect-father had built their homestead and saw for the first time Tow Hill, the gumdrop-shaped sailors' landmark which rises five hundred feet from low, surrounding country. The road wove behind Tow Hill, and soon we were out on the sandy beach. It was beautifully flat and lined with miles of driftwood. We drove as far as we dared go in the soft sand—which was about a quarter of a mile from the edge of trees at Rose Spit, one of the turning

Close-up of broken float containing Japanese "kita" marking on button near sealed end. This was found on the north shore of Graham Island near Tow Hill.

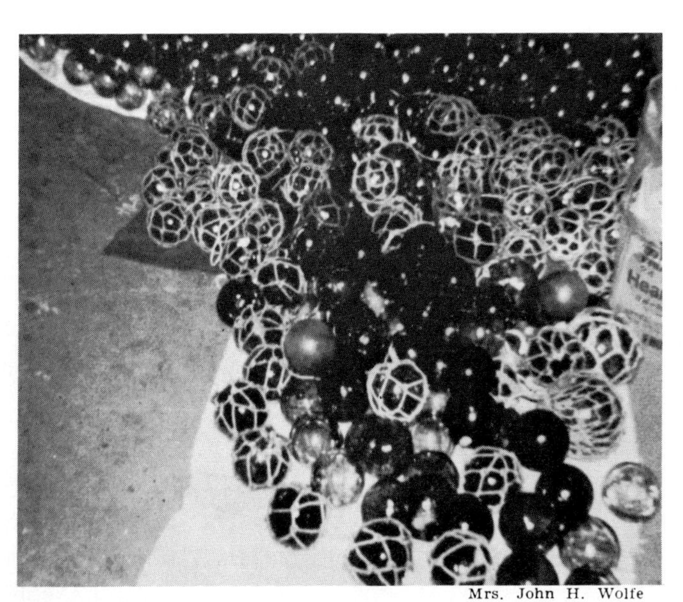

Mrs. John H. Wolfe

Part of the 721 St. Paul Island glass-float "find." Most of them appear to be salmon gillnet floats. Note the large number with nets still attached.

points of the great Kuroshio Current. Then Phil and I hiked across the spit to the east coast and into a driving southeast wind, while Norah and Elaine beachcombed in the opposite direction.

Wind had kicked up a foul surf. On this east coast of the spit, the beach is rather steep, mostly gravel, with the sand blanketing the driftwood, which is perhaps a hundred yards wide. We walked northeast along the beach for a while, then cut back across a myriad of wild strawberry plants to the west side, which was flatter. The surf was pounding and breaking out on the spit as far as the eye could see.

Walking back into a quartering wind was not easy; the sand bit into my face and my eyes watered. Here the storm had washed ashore numerous crab pots, several of which still retained good catches. We stopped to investigate the rusted wagon bed of an old steam tractor which had been cast aside from a gold-mining venture years ago. We also found a hand-made Indian oar. Out in the roaring surf a seal swam along effortlessly.

Meanwhile Norah and Elaine had beachcombed a good distance back toward Tow Hill. When we picked them up in the car, they had considerable loot. Shortly we detoured around a whale that had been beached at the edge of the drift-wood. At Yakan Point, Elaine and I got out and beach-combed. Here the road turned inland but would join the beach farther down. Elaine worked the beach high-tide mark while I searched the sand-dune edges and the higher grassy areas. Again we found more crab pots. Some rain came in turbulent gusts, but in general it fell in a mild manner. The two miles passed quite rapidly despite our slow and thorough search.

At White Creek we saw Phil waving from the top of a sand dune. It felt good to get back into their warm car. During the drive back to Masset we discussed shells and barnacles. After dinner we began an international bridge tourney, the Masset champions against the Mercer Islanders. Following an arduous competition the evening ended almost a tie. Jerry, the black Labrador, slept peacefully by our feet despite the storm outside, which by mid-evening had built up to gale proportions.

How very snug we were that night. Norah's toast and cocoa had concluded a most eventful day. Before going to sleep, as I thought over the day, I realized we had completed another goal, that of beachcombing Rose Spit, one of the great "turns" of the Black Stream. My theory seemed to have been right: we had beachcombed a dozen glass floats that day.

Alaska

In contrast to the commonly accepted image of a rock-bound Alaskan coastline, stretching some thirty-four hundred miles from Adak Island of the Aleutians to Duke Island south-west of Ketchikan—there are many beaches along this coast that provide excellent beachcombing for the glass-float collector. Kuroshio affects most of this shore area, often favoring places like Cape Spencer, Kodiak, and Yakutat. In the summer of 1936, members of an expedition to the Aleutians, studying prehistoric emigration from Asia to America, reported finding glass floats on the northwest shore of Kiska Island.

In September 1962, near Port Heiden, on the north side of the Alaska Peninsula, a fisherman rowed ashore and in a three-hour period retrieved over one thousand glass floats on a two-mile sandy beach. He said that westerly winds brought them in, and if he had looked farther up beyond the grass line he would have found thousands more. Because the Japanese fish near this area, sixty per cent of the gillnet floats found still have the cordage net attached. Apparently here in the region of the Bering Sea, the natives have placed no value on these floats until recently, when they learned they could use them in barter.

Trinity Island off Kodiak is another good beachcombing site; also Jones Bay in Cook Inlet. Floats have been so numerous at Cordova that a long path to the Coast Guard Station has been lined with them. Elsewhere these floats have been used in fences. With pangs of regret, I must relate that on three different occasions fishermen have told me that, at beaches where floats were plentiful, they would often throw them at one another, breaking them just to have something to do. One said that as a youngster he broke hundreds merely to smell the strange odor that was emitted.

In the Bering Sea west of Alaska is the Pribilof Island group, the isolated yet famous breeding grounds of fur-bearing seal. Mrs. John H. Wolfe, on St. Paul Island — which is the northerly island of this group—has related that the beachcombing of glass floats is excellent there, particularly after a steady storm condition. The Wolfes now have in their collection close to four hundred floats. In December 1964, two men in about three hours picked up 721 glass floats here, almost all of which were the 3¼-inch size. Another family, during the same period, found nearly one thousand. In that same weekend it was estimated that over two thousand were picked up on St. Paul Island alone.

Other items found on the beaches here include bottles, fishing floats of all types, soy-sauce buckets, large wooden mallets, bamboo poles, fish nets, and an occasional foundered whale. In the past, this island was a haven for walrus, and occasionally Mr. and Mrs. Wolfe still find the ivory tusks or parts of tusks from these walrus imbedded on the beaches or among the rocks. Beachcombing ivory on St. Paul Island is considered a real achievement by the local residents.

Hawaiian Islands

Japanese glass floats can be found in abundance here in the islands. Though Hawaii is not at one of the lucrative "turning points," it is just north of the westward-flowing North Equatorial Current. It is the "Kona" storms that bring in the floats. "Kona" is Polynesian for "leeward" and refers to storms from the opposite direction of the Northeast Trade Winds. These

Hawaii Visitors Bureau

Everything from bright-colored reef fish to glass fishing floats is being scouted by this group shown on a windward beach of Oahu Island, Hawaii.

storms occur most frequently during the months of October through April.

A recommended place to beachcomb is the Kaalualu Beach near Naalehu, toward the South Point of Hawaii Island. The clipping below from the Honolulu *Herald* of January 4, 1963, reports, under the title, "Hawaii's Top Sport is Hunting Fishing Floats":

> Glass ball hunting is becoming a popular sport in the Kau district of Hawaii with hunters coming from all over the island to take part in the sport. The attractive floats are often found resting on bagasse — washed sugar-cane trash that floats in along the island coast beaches. Finding a large, gleaming, aquamarine-colored ball resting in a nest of straw-colored bagasse is just as thrilling as finding a pearl in an oyster.
>
> Experienced Kau beachcombers who take the trip to Kaalualu beach, a few miles from Naalehu on the South Point road, make a hike which may range from one to four miles along the coast. "Hunters" feel the ideal time to find the glass balls is right after high tide. The winter months and stormy weather are very important, they feel, as to the amount of balls found. . . .
>
> Many floats begin their voyage to Hawaii from the northern Japanese Sea, drift past the Siberian coast, across the Bering Straits and down the Alaskan coast off the United States northwest coast of Washington and Oregon, then begin their southward and outward trip toward Hawaii. . . .

Wake Island

Two thousand one hundred miles west of Hawaii is the small V-shaped island of Wake, where according to reports, every day and at almost every hour someone finds a glass float. Mrs. Jessie Murphy, a beachcomber of Wake Island, tells this story:

> Our island is very narrow, so we all live practically on the beach. We don't consider this a danger. High tides

have on rare occasions dampened some of our floors, but we do not have tidal waves here. The tides themselves bring in very little except glass balls, and occasionally large logs which resemble poles. These logs lie on the beach until they are taken out again or are broken up. Empty bottles also wash ashore. Twice, notes have been found in them with strange addresses.

Floats come in all year around, but when there is a strong east wind blowing, we seem to find a greater number of them. September, October, March, and April are supposed to be the best months, according to some people —others disagree.

There are many glass-float collectors on the island: men, women, and children. We haven't any idea of the number of floats in the various collections but have been informed by a resident of fifteen years that, in the past, thousands of floats have been collected. The Wake Islanders use them for patio decorations and living room decorations. Many floor lamps have been made using the large glass balls. Some people put their glass balls on shelves, others in basket containers on tables or floors.

The Trents and Watsons have found over two hundred glass balls in the last five years. These range in size from a 48-inch circumference to the 3-inch size. The largest percentage picked up are the orange-sized ones and then the grapefruit-sized. Both families have discovered netted floats—six to eight each—ranging in size from a volleyball to a basketball. They have also found numerous rolling-pin types. Mrs. Trent has one of these about six inches long, with Japanese writing translated to mean "Big Boat on the Waves."

We find glass balls all along our beaches when out walking. However, sometimes for days we don't see any. This simply means that either someone has gone ahead of us or the glass balls are rolling up on another part of the beach. The rarest float on the island is a purple one about volleyball size. The next rarest are several large amber-colored ones. I believe my green one, fifty inches in circumference, is the only one this size picked up on the island. About five times as many little ones are picked up as large ones. One reason may be that many of the large ones are broken along our rocky coastline. We know, from broken pieces found, that occasionally a blue float comes in. All of us have three to six floats containing about an inch of water. These come in different sizes.

Very few of the floats are frosted. I've seen only two of them. Most of these are sea-green in color and clear. Netted floats are difficult to find now, and all of the floats are partly covered with a mossy scum. The larger ones usually have many barnacles clinging to them. Only a small percentage have identifying markings — so-called trademarks.

Wake Island—where the above beachcombing experiences occurred—is a spot in the central Pacific Ocean little larger than the airplane runway it contains. There, beachcombers in sun suits find Japanese glass balls. Other beachcombers, at Vancouver Island—clad in heavy woolen trousers and some five thousand miles downstream in the great Kuroshio Current—find similar floats, but a full year earlier. In other words, the floats that did not stop for the Vancouver Islanders might be found a year later by the Wake Islanders.

It is certain that very little is known about the distribution of lost glass fishing floats within the Pacific Ocean, but we do know where these floats come in by reports from beachcombers around the North Pacific Rim. These reports provide us with two additional pieces of information: there are a great number of beachcombers all around the Pacific, and they do look for glass floats. In other words, these highly traveled spheres get picked up and carried home, regardless of where around the Pacific Ocean the beachcomber resides.

The appeal of the lost glass ball is independent of the latitude and longitude of a particular sandy beach, and the color of the skin of the nearby beachcombing resident. The Aleut, the Indian, the North American, and the Polynesian — all search after a high tide. For those of us who walk miles and miles, only to come home empty handed, there is consolation in the experience of one Oregon housewife who said happily:

"I went out toward the cove one morning last week just as it became light. As I rounded the abutting rocks at the edge, there ahead of me were glass floats spread about the whole cove. There were so many, I didn't know which one to pick up first."

A friend of mine at the office once chided me by stating that, since all the places to go beachcombing were covered, perhaps I had purposely omitted one spot I didn't want anybody to know about. At the time, I had to admit that there was one place I didn't want to reveal, mainly because I had not yet explored that area; nor had anyone else, insofar as I could determine. But I do know this: once I take the helicopter trip into that relatively inaccessible beach, and search through the driftwood, and fly the treasure home—there will come to light another place to explore and another, and still another after that.

Floats found during a three-day beachcombing expedition to the Queen Charlottes. Vinemaple has been added to the large, broken float for an autumn-season centerpiece.

CHAPTER 10

Floats as Decorations

There are many decorative uses for beachcombed glass floats, either singly or in a group. As curios, they are conversation pieces; and, like a flower arrangement, a bonsai garden, or a piece of ornamental pottery, they can be placed almost anywhere within a room. They are also a subtle personal reminder of carefree vacation days . . . "Take note, the float is quite free, it has broken away from its net . . ."

A float displayed in proper light can be as beautiful as a soap bubble, showing the spectrum of color particularly where a separate light falls on its surface. To carry the comparison further: both bubble and ball are produced by blowing, both are spherical in form, both are smooth surfaced, and both have thin walls—but one has only a brief life, and the other will last indefinitely, forever lovely.

Oriental glass floats fit readily into the decor of both the traditional and contemporary homes; however, there are certain rules for their most effective display. To bring out the true color and texture of the Japanese glass float, it must be displayed where there is an abundance of light, especially in the case of darker-hued floats; the darker the glass, the more difficult it is to look through and find the real color tone.

Almost any piece of glassware, to be shown properly, requires special auxiliary lighting. The finest highlights will occur at one of the angles, depending on the shape of the par-

Three types of floats for display.

A group of seven glass floats displayed on the window seat in a corner of the family room of the Author's home. These floats were beachcombed on the outside of Vancouver Island, by helicopter.

ticular item. Steubenware and other high-quality glassware are thus skillfully exhibited. For night viewing, a float should be displayed as near as possible to the best light source in the room; for daylight observation, it should be near a window. In hours of darkness, a single glass float shows best when the light source is from below.

A group of floats at our Mercer Island home is arranged on a corner window seat with daylight from the east and south reflecting on and through the floats. I have found that window sills and window seats are prime spots for exhibiting a float collection. Plastic or glass shelves across windows will also do nicely.

Spherical glass shapes are often associated with light sources such as the common light bulb. When this glass shape is not light-generating, it becomes a light-rebound surface, especially in the case of smooth glass; and it will highlight some other light source from almost any direction. It is believed that many birds will avoid an outside lawn or garden area when a mirror-type ball is mounted there, as the sun will be reflected toward the bird's eyes regardless of where the bird may be. It is a matter of history that, in England, round-colored glass balls were hung in windows to discourage witches from entering. These hollow glass balls, generally called witch-balls, found their way to Massachusetts and were used as charms to ward off the evil eye.

Float Nets

Many people believe that a glass float has more value, character, and authenticity if it retains its original attaching net; thus many floats are found hanging on a wall or from an overhead beam. Too often, though, the net covers so much of the surface of the float that it is difficult to see the glass—let alone the color of the glass. The beauty of the float is mostly lost when it has no light to accentuate it. If a collec-

Display technique for Oriental glass floats that allows outside light to show the bubbles and color of the glass. Even the net-covered float casts reflected highlights. Note the glass collars under the floats.

An example of displaying a large glass net-covered float in such a manner that no light is allowed to pass through the glass. This is an inferior display technique.

tion is located indoors, the effect will be best if not more than ten per cent of the group have net coverings.

A friend of mine has a large float found in 1932 by fishermen in the Aleutians. It has the original covering of hemp-rope net. The rope is about a half inch in diameter and woven so tightly and close that the space between each stitch of rope is also about a half inch. This eighteen-inch float is hung in the corner of a porch in such a way you have to look up to the ceiling to see the float. Although I took the float down and held it to the light, I still could not discover its true color. It is, however, a most unusual porch conversation piece, and my beachcomber friend is to be commended for keeping the net intact—even though it does look like a big ball of rope. Reports indicate that only about one per cent of the floats beachcombed in Oregon and Washington still have their attach-nets intact.

Techniques for Displaying

Though there are many ways to present glass floats more effectively, most of them are displayed hanging in nets. The nets may be the original attaching cordage, used fishing gill-net material, or white-string artificial netting. Some curio shops add a white string to the net for hanging the float. The artificial netting generally overcomes the objection of not being able to see the color or texture of the glass. But, to fully appreciate the bubbles in the glass, the color, the shape, and any trademarks that might be imprinted, these Oriental floating treasures should be displayed unadorned.

Friends on Mercer Island have hung several large floats without nets from the ceiling of their living room on single-filament nylon fishing lines. From a short distance these appear literally suspended in mid-air.

Imitation glass floats—manufactured in Japan—serve for decorative purposes. They can be purchased in stores handling garden supplies or artware, and are usually recognized by their extremely clear glass and rich coloring. Frequently they come with a thin, string-like mesh net which readily reveals the color and texture of the float. The Japanese also produce attractive imitation square-glass floats of various sizes and in such colors as deep gold, wine, and clear brown. These floats are best for group displays, though I have occasionally seen them used as light reflectors in outdoor ponds.

I was once given a glass-float cube, two inches on a side and dusty lavender in color—but sealed with the regular collar button of commercial Japanese fishing floats. The flattened sides of the cube were pulled slightly inward, probably because of their "setting" during cooling. It is difficult to imagine any commercial fishing use for this shape of float, but it is certainly a decorative addition to flower arrangements and a popular conversation piece.

A large home on Orcas Island, in the San Juans, uses glass floats for the ceiling of the recreation room, and the outline of the floats is the shape of the island. Small floats are arranged in such a way that, at the flick of a switch, the overhead light comes through the green glass balls. Several larger lighted floats, placed strategically, provide additional light. The total effect is very unusual and very pleasing.

A shadowbox is an artistic way to display a float collection as well as other glassware. It should have three glass shelves, a mirror behind, and preferably glass doors; light should come from below and above. This arrangement presents double images of each item and accentuates color tones. It is particularly effective for roller and cylindrical floats and those with special markings or other unusual characteristics.

There is reason to display floats in a group of different sizes. A single small one has little to present, except close up, and a single large one doesn't look as large as it really is; but a group of varying sizes will show each float in proper perspective.

Another attractive home display can be achieved by using driftwood or sea shells with the floats. Some driftwood with knotholes offers a unique arrangement: just fit a small float snugly into the knothole.

Some glass floats are more than decorations, they are much-prized trophies. Take the case of the California lady tourist who stopped her car at a view spot along the Oregon Coast Highway. Having in the past searched unsuccessfully for glass floats, she was surprised and elated to see one about to be beached in the surf below. Observing that the float was on its way in, she determined to go after it. With no concern for life or limb, she worked her way down the rocky, two-hundred-foot cliff, splashed out into three feet of water, waited for the float to come within reach, and picked it up. Then, clutching it tightly, she climbed laboriously back to her car. The ruined shoes and snagged nylons—let alone the bruised knees and scratched arms—were all forgotten when, two weeks later, she placed the dull-blue four-inch sphere on the mantle above her fireplace.

One of the most attractive landscaped yards using glass floats is that of Bill and Myrtle Saxon, on Cox Bay near Tofino. They have placed about eighteen large floats and many smaller ones at the base of wind-swept spruce trees and moss-covered stumps. The arrangement is altogether pleasing and serene; the floats seem to have rolled right out of the nearby surf, up into their predestined proper places in the Saxons' waterfront yard.

A neighbor on Mercer Island has a fine collection of glass floats in his garden; the floats are placed in groups among plantings, weathered stumps, and atop a concrete bird bath. The collection came almost entirely from a single trip to the Lake Ozette Pacific Beach during the "Big Blow" (ninety-knot rain and wind storm) of March 1960. A group that had hiked into the Cape Alava area just ahead of him also did very well;

they had camped out in a snowstorm in order to be first on the beach that morning to pick off the big ones.

The same neighbor tells about a business trip he made to Yakutat, Alaska. He was driven by car about fifteen miles down the beach from the Coast Guard Station, and along the way saw hundreds of glass floats just waiting to go out on the next high tide. He was told that, as a diversion, some of the Coast Guard personnel would pick up a few, pack them in popcorn, and mail them to friends and relatives.

At a home near the crab cannery at Tofino, there is a four-inch roller glass float being used as an insulator for radio antennae. To one end is tied the radio antennae, and to the other a short line to a pole holding the wire. Eight miles south, near the Tofino Aerodrome, several large floats are placed at random in a pasture where cattle and sheep are kept. I stopped to inquire, but the owner was not at home, so I drove away still wondering why large floats were placed like that.

In Tofino, several dark-brown, machine-made floats adorn the railing and banister of still another home. Apparently this owner prefers to display the rarer brown floats rather than the green Oriental ones from the nearby beaches. Many restaurants—and they need not necessarily be close to the water—feature glass floats in their decor. Ivar Haglund, well-known Seattle restaurateur, has used them effectively in the interior decor of his waterfront restaurant, Ivar's Acres of Clams. There on display are many fine examples of floats of all sizes, colors, and markings. The most unusual one is built into the wall like a porthole, just to the right of the entry door: a twelve-incher containing a filament-type spindle. I was assured that it is so well built into the wall that no harm will come to it. Ivar's Fifth Avenue restaurant artfully employed various-sized floats in the ceiling in such a manner that light was reflected through them.

This is how he got his floats: A man and his wife in a camper from Oregon called on him and tried to interest him

British Columbia Government Photo

Trader Vic's Restaurant, Bay Shore Inn, Vancouver, B. C., showing use of Japanese floats.

A ten-inch glass float showing the separate trademark button. This marking is readily identified with its TY arrangement, although the Japanese symbol is read from the inverted position. See No. 25, page 213.

in buying their collection of floats. Ivar wasn't really interested, in fact not at all. They then told him their hard-luck story of trying to sell them first in Tacoma and then all along the Seattle waterfront; they wanted to sell them in order to pay for their two weeks' vacation. The collection was the result of five years of spare-time beachcombing on a spit near Coos Bay, Oregon. Finally Ivar consented to go out and look at the collection in the back of their camper. There were more than three hundred. A deal started and soon was consummated. The couple were happy with the cash for groceries and gasoline and Ivar was happy too. Where else could you buy the results of five years of beachcombing on a desolate Oregon sand spit, and decorations for your new restaurant—all for $275?

Other restaurants also employ glass floats extensively for decorations. They fit in admirably with the Polynesian decor of the Trader Vic restaurants: large floats of excellent color hang from the ceiling beams or are displayed among huge South Sea shells and carvings.

When placing floats on a table or stand, there is one cardinal rule: always use a stabilizing collar for each float so it will not roll away. This collar can be a small plastic ring, but a metal washer will do, and a glass furniture coaster is ideal. These collars also preserve the original grouping arrangement of different-sized floats. Three of our prized beachcombed items got broken before we learned. And when did they get broken? During dusting. To break a glass float at home—after it has traveled perhaps ten thousand miles and you have gone to great effort to find it—can be a real catastrophe.

Such a catastrophe occurred on March 14, 1961, off the north jetty at Cape Disappointment near the mouth of the Columbia River. During a hard rainstorm and right at the semi-darkness of daybreak when nobody else was around, a friend found three floats. Not having a knapsack or back pack, he faced the problem of transporting them. The largest was around twenty inches in diameter, the second close to fifteen, and the third about five. It was unthinkable to leave any of the three behind for someone else to grab, and unfortunately there was no good place to hide them. My friend's solution was to carry the two smaller ones, one under each arm, and roll the big one along the shore with his foot.

All went well for about a hundred yards when he needed to get a better grip on the fifteen-incher. In so doing, the five-incher accidentally slipped out of his arm and fell. Yes, right down and through the big one, leaving a large jagged hole. I am told that the sound emitted from this shattering glass was indescribable, momentarily drowning out the roar of the ocean itself. My friend attempted to salvage the big shell, holding it by the opening; but as he was walking back over the drift-wood, his handhold broke and the remainder of the float fell and shattered. For a long time he kept a jagged portion to show its size.

Painted Floats

An unusual way to increase the artistic appeal of larger glass floats is by painting pictures on them. Because I had heard of Indians doing landscapes and the like on some of these floats, I decided to interview Mrs. Hayes of Long Beach, Vancouver Island, who was reported to do this craftwork. As I walked down the path leading to her home at the north end of the beach, I was greeted by the following formidable sign:

> Notice: This is an Indian Reserve. Any person who trespasses on an Indian Reserve is guilty of an offense and is liable on summary conviction to a fine not exceeding fifty dollars or imprisonment for a term not exceeding one month, or to both fine and imprisonment.
>
> Director of Indian Affairs
> Department of Citizenship and Immigration
> Indian Affairs Branch
> Ottawa

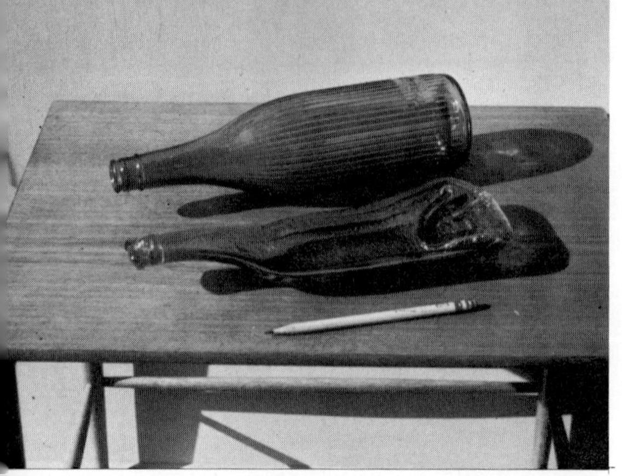

A beachcombed saki bottle like this wa
melted down into a celery dish by ceram
ist Jeanne L. Adams of Seattle.

Ruth Burton shows artwork she did o
glass floats found on the Queen Cha
lottes.

Two ashtrays made from pieces of broke
glass fishing floats by Betty Brown of Ker
Washington. The larger tray was fus
from several pieces. The wide, flat bubb
adds the artistic touch.

I wondered about possible consequences with the Mounted Police, but decided to go ahead; I reasoned that my mission was innocent enough. I soon found Mr. Hayes, who was changing a tire on their car, and I explained the reason for my trespass. Without a word he took me to their small green-roofed cabin and pointed to Mrs. Hayes who was sorting sword ferns. Again I explained my mission. She began by telling me that they had just returned from picking ferns which they were packaging for shipment to florists in eastern Canada. Then she brought out some glass floats they had beachcombed from Schooner Bay. Continuous pictures had been painted clear around the circumference of each ball. The paintings were authentic art craft, made with Indian-formula paints which she herself had mixed. The scenes were also of Indian origin and depicted the sea with the rocky shores and the trees of this area; some contained Indian words. I believe this particular craft is limited to the local Indians.

Mrs. Hayes explained that her husband was a retired fisherman of the Clayoquot Tribe; and that before retirement they would go out during the fishing season and live for weeks on their boat or on a nearby beach. Then they would move back to their big house at Opitsat, an Indian village on Meares Island. She talked of the storms that destroyed so much Japanese fishing; how fishing gear and even whole boats were lost; and how large portions of Japanese nets, with their floats still attached, washed up almost in front of their home. Soon I thanked her for the interview and went outside. Mr. Hayes was still puttering with the car. I thanked him also, and he nodded. I looked for the Indian Reservation sign as I walked back to my car, but somehow I missed it.

Glass Floats as Lamp Bases

Many people like to make table lamps using medium-size floats for the bases. This involves drilling the float in two places in order to slip it down over the metal pipe containing

These two large Japanese floats with hemp-rope attaching nets were found on Alaskan waters. Of special interest are the wooden nametags still attached. It is believed these floats are used in the Alaska king crab fishery.

the wiring. Some lamps are made with several drilled floats stacked on top of one another. However, most of the completed lamps I have seen consist of a single eight- or ten-inch float resting on a wooden base. Another attractive design for a lamp involves using a length of copper tubing which, when coiled, serves as the base, then emerges into a stem to encircle and hold the single float and contain the electric receptacle.

Drilling Floats

Drilling glass floats is a special art mastered by only a few. At Long Beach, Washington, there are several persons who do this, but generally they do not guarantee their work; the breakage is high. With the pounding that some floats must take rolling across cobblestone gravel during beaching, many small fractures can occur but go unnoticed until the drilling process enlarges them, often disastrously.

For holes up to one-half inch, it is best to use a drill bit made from a piece of copper tubing that has teeth cut in one end. In the drilling operation, an abrasive such as carborundum in granular or paste form is used on the glass, allowing the tubing tool to grind its circular opening. The abrasive needs frequent replacement. A few drops of water provide the drilling lubricant.

A simple and sturdy fish aquarium can be made by cutting a four-inch hole in the top of a large float. For years one of these was prominently displayed in a window on the main street of Long Beach, Washington. However, I have never had the courage to select one of our own large floats and experiment with it in this fashion.

Wooden Nametags

An outstanding decoration is one of the large net-covered glass floats that still contains a wooden nametag. Floats with nametags are considered collectors' items. These tags are

David Davis

Verna S

Left: Floats suspended by clear plastic fishing line in the home of David Davis on Mercer Island. Note the interesting shadows cast against the wall. *Right*: What to do with your beachcombing loot is demonstrated in this attractive display of floats, fish lures, and "what have you." The arrangement is at the Farrell Blum home, north of Lincoln City, Oregon.

about two inches wide by four inches long and one-quarter inch thick and are sometimes found attached to one of the rope loops. It is believed these wooden markers are found on floats that have been used in Alaskan waters in the king-crab fishery. I have not learned of any being beachcombed elsewhere around the North Pacific Rim.

I once took a copy of the markings found on a nametag to a Japanese friend for translation, since he had done this sort of thing for me before. He returned my sheet of paper with the following: "March 27, 1942, Send Help. Outnumbered." I looked at him with complete amazement. He then smiled and handed me a second slip which read: "March 27, 1942, *Ju-Ji-Maru*"—probably the date the net was set and the name of the ship. The joke was on me.

If a glass float is found either with or without a net but with barnacles still attached, and it is desired to retain its total appearance, a method which appears quite successful has been used by Mrs. Ray Wilson of the Wecoma Beach district of Lincoln City, Oregon. The complete float is submerged overnight in a ten per cent solution of formaldehyde. After drying, several coats of plastic are sprayed on. The result is attractive and lasting.

Almost all who visit or vacation at the Pacific Ocean beaches want to take home a reminder of this greatest of oceans. Often a shell or agate suffices, but for those of us who are looking for a symbol, something to project our thinking beyond just the sandy beach and crashing waves, the glass float is a natural choice. It becomes a conversation piece, a springboard for talks of beach trips, an object of wonderment. But, to attract attention, the float needs a stage and a spotlight to tell its story. It is for each of us to display it according to our personal wishes and ingenuity.

Verna Slane Ph

David Coffee's collection of glass floats, found along the 20-Miracle-Mile beaches near Lincoln City, includes all shapes and sizes.

CHAPTER 11

Collectors and Collections

There are more collectors of Oriental floats than collections. Some collect to sell, others for fun, a few to hoard; but there are relatively few satisfied collectors. They are great horse traders, always with an eye toward something they want. Some specialize in color, others in markings. Some are always on the lookout for an additional colored float. Others may collect just oddities in shapes, like rollers. A collector in Oregon is reported to own eight large rollers, about four inches in diameter—quite a rare collection.

Some believe that, in time, these glass spheres will become rare. However, there is now good evidence that millions of glass floats are still spread about the Pacific Ocean, and that this reserve is increasing, not decreasing. It is true that with continuing technological developments, the day of the glass fishing float may come to an end but, according to my studies, attic hoarding will not pay off very fast.

The individual who hoards floats against the future may have to wait until the year 2145 A.D. before the Pacific Ocean has given up the last of this floating treasure from Japan. Statistically speaking, if the manufacture of glass floats ceased this very year, it would take approximately twenty years to diminish the North Pacific float reserve by half—assuming the same incidence of storm and wind as has occurred these past fifty years. And, even though the larger fishing operations will employ more and more mechanization, there will still be the smaller hand-operated seining; Japanese fishermen will, for certain, continue to use some glass floats.

I believe they will be the number-one prize of the American beachcomber for many, many years. So, if you must hoard these treasures in your attic, plan to store them for a long time.

Every float found has an inherent story but much of it must remain hidden. An exception to this is the case of some Danish-made floats displayed at the Seattle World's Fair. Identification had been added to one side of each of these eight-inch green floats; but few floats are so marked. To try to determine the location and date of finding for each individual float is a chore few care to bother with. Besides, the place and date found is only a small part of the total story. The origin and travels are needed for the entire identification.

Marking Floats

Etching or marking of floats can be done in several ways. The first and recommended way is with a diamond-marking pencil, obtainable from a laboratory supply house for about two dollars. Practice a bit on some spare glass, then mark your float. A second method, using paraffin and hydrofluoric acid, is not recommended for the amateur collector unless he is skilled in chemical laboratory practices. Nor is an electric vibrating marking tool advised, since the float might fracture or blow in. Marking floats by painting on dates and so forth can be easily done, but it is not very satisfactory; it detracts from the appearance of the float.

Collections

One of the most colorful and artfully displayed private collections belongs to the Ted Blodgetts of Seaview, Washington. Their collection consists of quality items selected from ten years of beachcombing. To display them properly they have had to limit the number of large ones to twelve, and these are insured for a thousand dollars. They have remodeled their

Left: Imprinted on this beachcombed bottle was "U.S. Department of Commerce, Coast and Geodetic Survey." Inside was a card to be mailed in for a reward. *Right*: A novel way to identify floats that have a special story.

Hjalmar Brenna

Glass bottles from many countries are also picked up on Pacific beaches. The one at center left apparently traveled all the way from New Zealand.

home to accommodate their handsome collection, which is arranged on shelves against large windows. When the sun shines through the panes, the whole room sparkles with color. They have worn out three Jeeps driving on the beach, for most of the driving was done in four-wheel-drive in low gear in the sand. These beachcombing forays are reflected elsewhere in their home, which is paneled with beautifully finished pieces of mahogany, teak, holly, and myrtle—all collected from near their front door.

Harry Tinker of Long Beach, Washington, has a collection of 640 floats displayed on racks in his back yard. With more than fifty years of beachcombing and shipwreck salvage behind him, he says that Japanese glass floats still create more excitement than any other beachcombed items. When these floats were first found on this beach, he says no one had the slightest idea what they were or where they came from. He tells of one costly experience while driving the beach: he had been caught in a high tide because of using a tide table of the right day but the wrong year. While grinding through the sand, the clutch of the truck gave out, and a big wave knocked the truck over on its side. Shortly thereafter ten men from the Coast Guard Station were able to rock it over onto its wheels, but by then the sand in the surf had worked into every nook and passage within the engine. Dismantling and cleaning each part of the engine cost as much as a new truck.

A collection of representative sizes, shapes, and colors in floats belongs to Mrs. Virginia Murray Bone of Seattle. All of her five hundred floats were gathered prior to 1942, and there are many choice and rare items among them.

Since a collection of much more than a few dozen floats constitutes a space problem, the determined collector either relegates the remainder to the garage or he begins giving them away. The front-room collection is usually show-sample rather than a total display.

Wayne R. O'Neil

Mr. Tinker and his glass float collection of more than 600 items, all picked up at Long Beach, Washington.

Wayne R. O'Neil

Gilbert Lind of Seaview, Washington, and his catch from one four-hour tide.

As we traveled south along the Oregon Coast recently, I was able to locate several collections containing as many as one hundred floats. Mrs. Robert Wise of Nehalem has more than one hundred in her collection, the result of over four years of beachcombing the Nehalem Spit. Her floats include an eleven-inch roller, a float frosted inside, and two of a brilliant blue color. Most of the floats she finds have some sort of marine growth on them.

Not all glass floats are used out on the high seas for the fishing of king crab, tuna, or salmon. In quiet Toba Bay, off the Pacific Ocean side of Honshu Island north of Nagoya, Japan, the tourist who stays at the Toba Hotel International may look down upon acres of oyster baskets assigned to pearl cultivation—all buoyed with glass floats. Our neighbor wrote: "We are looking down from this hotel on about nine hundred glass balls, approximately sixteen inches in diameter. . . . We haven't figured out yet how to bring one home."

We all learn from experience and we often learn best from firsthand experiences. The following story concerning Bud is not about collecting glass floats, but it is directed to beachcombers:

It seems when Bud was a small boy, his father one time said, "Son, you should carry a good pocketknife." And so Bud always did. It may not have been a necessity, but it had been a handy item to have on one occasion or another. He can't exactly explain why he always carried it, other than to have obeyed his father.

One day Bud said to his own son, who was then about eleven, "Son, you should carry a pocketknife." To which the young lad queried, "Why, dad?"

"Well, I guess I can't give you a good reason," Bud said. And so, in typical, modern father-son relationship, the son proceeded along his merry way without having taken to himself a knife.

Seattle Tim

Seattle halibut fishermen display Japanese fishing gear found in the closed waters of Bristol Bay.

Months later, at a family social gathering, one of the guests was the survivor of an airliner emergency landing off the Oregon Coast. She answered many questions about her thrilling adventure. She told how everything had worked out to perfection, how the pilot had wired ahead for assistance, how flying boats had arrived and were standing by, how the water was smooth and a perfect landing was made by the airline captain, and how the life rafts were inflated and the passengers were in them well before the plane started to fill with water and sink. She also said that, to keep each raft close to the door for loading, the stewardess had tied the nylon line to the door hinge.

About the time the last raft was filled with people, the water was up into the cabin, and shortly the plane started to sink. Too late it was noticed that the raft was still tied to the airplane; in seconds it started to be pulled under by the sinking plane. The narrator told of her own frantic thoughts as the water started to pour in over the end of the raft. Then, suddenly, seemingly out of nowhere, an elderly man, whom she hadn't even noticed before, reached into his pocket, took out a small knife, and cut the cord . . . . Bud said that his son, who had listened closely to the story, now carries a pocketknife.

Ralph McGough tells a strange story that happened near Oysterville, Washington. He had watched a large float as it rode toward the beach through a heavy surf at high tide. Although it was difficult for him to estimate its size, his practiced eye told him it was a real prize, maybe eighteen inches in diameter. He waded into the water, anticipating its retrieval, as he had done so many times before.

The float bobbed toward the final three waves. He watched it ride up and start down as the first wave crested and broke. The second wave built up and crested, but the float was nowhere to be seen. He waited but the float had vanished. Nothing came out of the second wave nor the final one. Since he

W. Lyle Kirk

Part of Dan F. Guffey's collection of about one hundred Japanese glass fishing floats found at Oceanside, Oregon, during the winter of 1960-61. Most of them appear to be long-line tuna floats as evidenced by the one that still has its attaching net.

wanted to find out what had happened, he marked the place with a stick in the dry sand.

Later, after the tide had gone out, he returned to where he had left his marker and checked the broad, sandy beach for some clue or evidence; but the sand was bare. There was no rock, no broken glass, nothing to explain the mysterious disappearance of the huge float. The Pacific had apparently reclaimed it. Possibly when rolling from the wave crest, the float had been driven into a shallow trough and had struck the hard sand and shattered.

Those who pursue the Oriental glass float are, generally speaking, of a stern breed: they don't mind braving the elements and they occasionally risk life and limb seeking their prizes. In short, they are members of that portion of the population who enjoy doing something a little bit different from what the next-door neighbor does. For example, I have a neighbor who collects guns, and we get along fine. He buries himself among his catalogs, and I retreat to the outer beaches.

Once beachcombing gets into your blood, chances are good that the lure of it will linger. A guest was telling about the glass float harvest off the Oregon Coast; how at times they came in great quantity; and how he would get up at night to go look for them. At the climax of his story he recalled how, with so many big ones coming in, he didn't even bother with the little ones. At the conclusion of the story, I asked a single question:

"Did you ever walk past a small glass float without picking it up?"

There was a long pause before the answer came in small letters, "No."

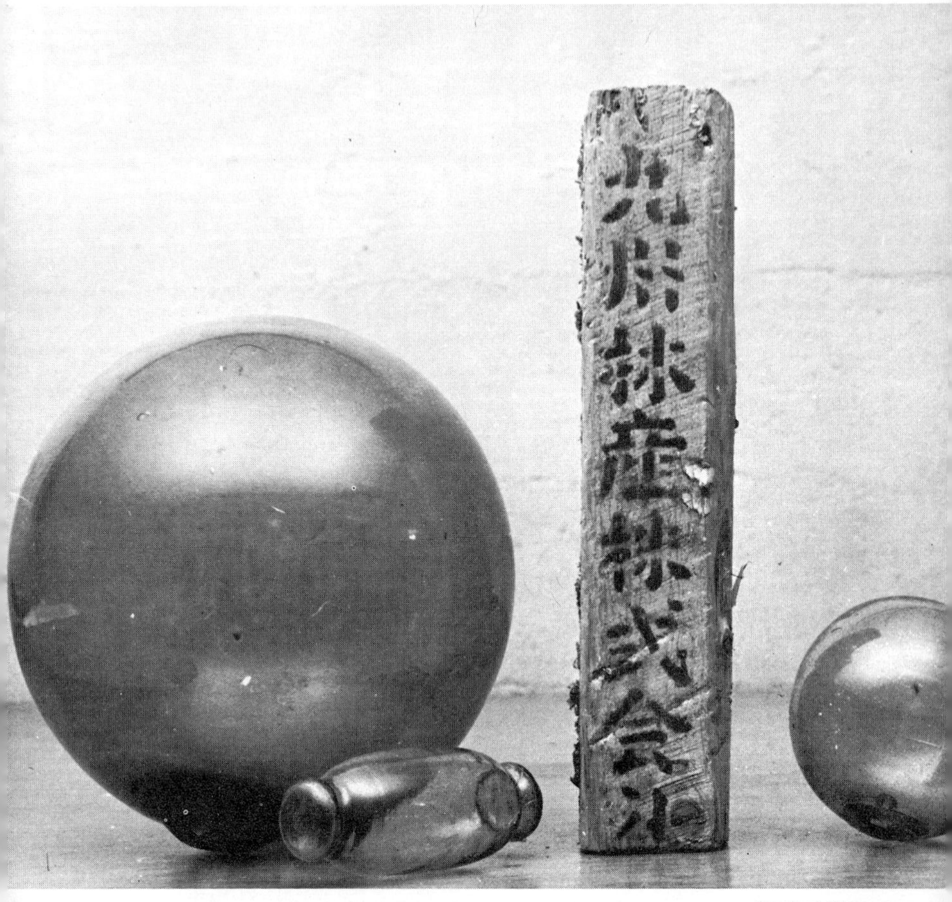

Burford Wilkerson

Oriental treasure for the Pacific Coast beachcomber.

CHAPTER 12

A Summing-Up

The after-dinner discussion following a satisfying day of beachcombing is buoyant, and a bit poetic. There is rarely mention of the many miles of travel to and from the beach, and the many miles of hiking for a few cents' worth of Oriental glassware. The talk usually keeps to a light vein and deals with experiencing nature's ever-present factors of tide, wind, and sky. I will maybe tell about the offshore wind which delayed the huge waves of the surf from breaking until the white plumes were torn from the top. My wife will perhaps describe the new shells or barnacles she found at the high-tide line.

Then will come stories of unusual storms and what was found on the beach afterward. We avoid any justification for the pursuit of the continually elusive glass float. It is but a part of a much larger influence; we are slowly being captured by the serenity of the sea. After periods of active urban living we keep coming back to the beach. It beckons us much as the sirens of yesteryear beckoned Ulysses' sailors.

The beach is our vacation home, and we try to organize our trips to include the winter and spring storm periods. There at the beach we straighten out our outlooks. We leave disturbing influences far behind; the clean salt air clears away the old worries. I use a pocket notebook to catch some of the ideas that evolve during our beach hikes. When I am moved to wonder, I stop and write the words that come. My notebook now contains all sorts of thoughts; some of them I have related here.

In retrospect, these beach visits seem to fill our special need to get away from what we consider our daily grind. I recommend to others this communion with the sea, and beachcombing for whatever you see fit. Our choice is Japanese glass floats; but it could be shells, driftwood, or marine hardware. If you also choose to search for glass floats, be prepared to find other treasures of considerably different value.

You don't have to spend much time at the beach to experience the exacting nature of the sea. In a storm it is cruel, ruthless, and demanding. The slewing about of massive pieces of driftwood at high tide during a bad storm has to be seen and heard to properly understand the sea's awesome power.

Yet after a storm the ocean may suddenly become gentle and tranquil. It was that way in August 1963, when Elaine and I went to Whidbey Island to spend the day at our cabin. After I had taken a brief sojourn on the cabin roof to cement some loose chimney bricks that were parting company after many years of mortared bliss—we took the four-mile hike along the beach to Indian Point.

The sun was bright, mountain peaks were just visible above a low haze, salmon fishermen were gathered off the north point of Gedney Island, and a mild, cool breeze drifted in from somewhere. Seagulls were congregated on the warm sand flats, where the tide had left choice morsels of whatever seagulls like to eat for a mid-morning breakfast. As we walked toward their domain, they took flight. Apparently we had invaded their privacy, so they moved farther down the beach. In the sand, about halfway in our journey, we saw tracks of both raccoon and deer.

After our long walk down and back from Indian Point, we took inventory of our loot for the morning. It included a ten-foot ladder which we had carried most of the way back on our shoulders, a white nylon pocket comb in good condition, an old wagon-harness D-fitting (which will be used in our boat sling), a candle which will cheer us some night when the power goes off, a beautiful long white feather, and an oarlock fitting still bolted to the wooden side of some rowboat which had been smashed to pieces. True, we didn't find any glass floats or interesting bottles on this trip, but I wouldn't trade our morning walk and our beachcombed bits for a whole week-end as the guest of an Indian maharaja.

We had the luxury of a quiet beach all to ourselves. What better way to spend a Saturday morning, after a busy week at the office, than to shed one's cares, hike down the sandy beach, and play this casual game of grab-bag at the high-tide mark? An early morning high tide had left a whole new array of bark, sea grass, broken branches, and all sorts of interesting things for us to search through. For only a short ferry ride from Mukilteo on the mainland to the sandy beaches of Whidbey, we were transformed from city dwellers to carefree beachcombers.

Although beachcombing any ocean shore is satisfying, we particularly enjoy combing small uninhabited islands. This does not mean that we have to go to some faraway region. Within a single day's travel from Mercer Island there are some excellent places. An ideal place is an island that has an outside beach exposed to prevailing storms, with no means for safe anchoring, but is accessible on the lee side by boat. Most of the sea-borne debris is cast up on these shores during storms involving considerable rain and wind, and this helps keep people away. If the island has a good harbor, fishermen will stop there more often to anchor during a storm.

Some of the small islands along the shores of outer Vançouver Island seem to hold no particular interest for the Clayoquot Indians. The fact that certain of these islands contain tribal burial grounds has a great deal to do with the Indians' avoiding these areas. They bury their dead away from the villages on a separate island in order to isolate the accompanying evil spirits. Besides, the access to these outer islands is usually difficult, requiring rock climbing or brush cutting to get into; and there is no ready crop of fish or lumber for the Indians to turn into cash.

We have beachcombed shores that bear no evidence of having been visited for weeks, even months, by any living being, other than migrant wildlife which occasionally may stop by. It is quite a sight to visit such a beach after a storm, but it is an even greater experience to be there during the storm.

Little effort is required for us to be at the right beach in the right storm. Elaine can be organized to leave in less than an hour, and I usually double-check the weather reports to confirm that the storm is heading to the target area. Then we toss the waterproof rain gear into the back of our vintage station wagon and start out.

Much too often, though, when a full-fledged storm is brewing off the coast of Vancouver Island, I find I am involved in a military airplane sales-proposal and can't get away to participate in the prime beachcombing. I am now convinced there is some relationship between Alaska gyral storm movements and airplane sales-proposal due-dates. It seems as if our project people are all too often in the final crucial days before delivery deadline when a beautiful storm kicks into the Queen Charlottes or Vancouver Island. If I ever retire, one of the things I want to do is go up to that outer beach on Flores Island and leisurely wait for a storm to arrive.

Transportation at these thinly populated coastal areas is by small boat. Because successful beachcombing is a matter of timing, we take along our outboard boat, motor, and emergency equipment. All that is necessary to impress one with the need for tools, fresh water, tarp, food, and blankets—in the event of a mechanical interruption or navigational error—is to spend a night, unprotected, out on a cold, rain-swept shore.

Some of our beachcombing ventures of course fall short of being completely successful. The over-riding enthusiasm of

After an early morning beachcombing trip with the outboard boat along the shores of Whidbey Island, the Author's wife carries a box across the tide flats to the cabin, while he unloads fireplace wood. Her treasure was converted into more shelf space for the kitchen.

Looking across Russell Channel toward Catface Mountain Range, Vancouver Island. . . from Flores Island, where beachcombing is ideal.

getting an expedition under way has resulted in many humorous muddles. Running out of money up in British Columbia isn't so bad, and forgetting to bring hiking shoes is all par for the course; but to expect our Canadian neighbors on Vancouver Island to observe the same national holidays has left us stranded on several occasions. Then, the decision of whether to bring the cameras along has often turned out wrong. During a driving rain, we once decided to leave the cameras in the car, only to come on a choice shot in bright sunlight two hours later. Sometimes when we tote the cameras all afternoon, good subjects seem to disappear. I now carry both cameras regardless of weather conditions and weight. Being able to record some special event overshadows whatever labor is involved.

There was the trip when we arrived at the end of the beach at lunchtime. The main course and the only item on the menu was our emergency ration of crackers and one can of sardines. To make matters worse, when we prepared to open the tin can, the little key required to twist off the lid was missing and we had to beachcomb for a substitute. Fortunately I was able to work loose a nail from a plank, and with the aid of a rock, I managed to pry open a corner in such a fashion as to be able to extract canape-sized shreds of sardines. Naturally there was a certain amount of sand from the rock and of rust from the nail that ended up on the crackers; but after that particular hike it didn't seem to matter.

For the Pacific Northwest beachcomber, an aura of mystery and romance surrounds the floating glass bubbles; but beyond their usefulness they apparently have no special significance to either the Japanese glassblower or to his fisherman customer. These industrious people are so involved in eking out a living for themselves and their fish-diet nation that they have little time or energy left to attach any special emotion to the product of their craft. To them these floats must be effective, and they cannot cost very much. The Japanese people traditionally have had to work hard, and with limited resources, to provide the necessities of life, so resourcefulness to them is synonymous with existence. Similarly, America's commercial fisherman usually has no artistic attachment to his fishing gear, which he too may soon lose, all in the day's work. Even the salvage value of lost fishing gear is small. If you have ever taken the trouble to repair a gillnet which has been washed up on a beach, you will know what a time-consuming job that is.

So, the often beautiful floats are only a part of the Oriental craftsman's tools while he earns an austere living. In the course of his trade, he will unhappily lose some of his floats to the sea, whereupon Kuroshio will send them across the Pacific Ocean to land on distant foreign shores: maybe Kodiak, maybe the Queen Charlotte Islands, maybe Whidbey Island.

To the beachcomber on the faraway shores it is quite another matter. As he stalks these glass treasures through the driftwood along some sandy beach and finally spies one among some bark and seaweed, his first reaction may be wonderment at how far it has traveled, how it was lost, and where it was made; he will likely have little concern for the unlucky fisherman. Here confronting him is man-made evidence of that mysterious Orient located far out beyond the horizon.

Not all that is found by beachcombers necessarily floats in on the beach. Sometimes objects so dense that they cannot float will wash up from the action of underwater currents and the surf. The heaviest non-floating piece of beachcombing treasure of this type brought to my attention resulted from an experience of a friend in September 1961, near Westport, Washington. He was searching the drift along the shore, about one-quarter mile north of the road entrance to Twin Harbor State Park, when he stumbled onto a half-buried piece of metal, molten in appearance and about the size and shape of a large summer squash.

David T. Davis

The beachcombing Wood family—Amos, Francie, Yuriko, Nancy, Elaine, and Dick.

It was crystalline on one side and had a number of holes or passages running throughout. He wrestled it for a short distance, but soon found that carrying this eighty-pound blob had become a first-class chore. He then dropped it by a big stump, intending somehow to get it to his car later. Well, with the press of other matters, he didn't get back to do this. It is probably still there, perhaps covered by the sand. As this friend is an artist, he later made a color sketch of his find and presented it to an aerospace expert. The verdict soon came back: the blob was likely a meteorite.

There are many consolation prizes for those who look for glass floats but just don't happen to find any. There are always the tide pools to look into and wonder at the marine life there. A friend once caught bare-handed an eight-pound silver salmon that was splashing about in such a small pool after the tide had gone out. There is the pleasure of being at water's edge when the smelt are running. Facing the shore as the water is rushing back, you can hold your hands in the shallows and feel the small fish running through your reaching fingers. This is not only for small youngsters. . . .

There are starfish to find; these can be preserved as decorations by boiling them for twenty minutes in salt water. There are the shells to bring home, and the memories. The beachcomber, though, will find something more important than all these. As he pursues the elusive Oriental floats along miles of driftwood shores, he will come close to an element so powerful yet serene that it will temper his entire philosophy of life—as it has ours.

No. 85

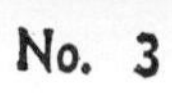

No. 3

No. 25

CHAPTER 13

Trademarks and Imprints

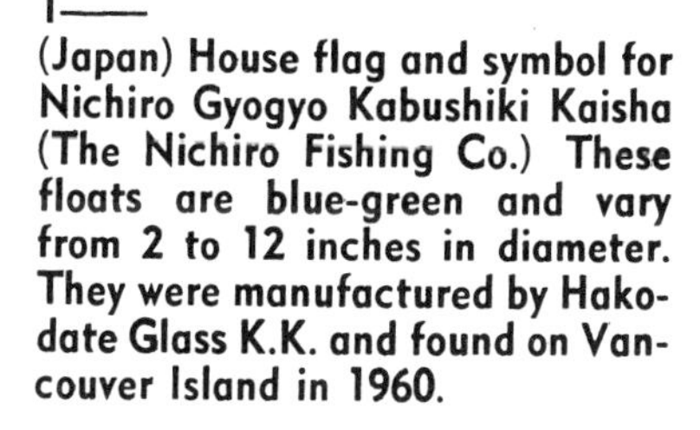

1——

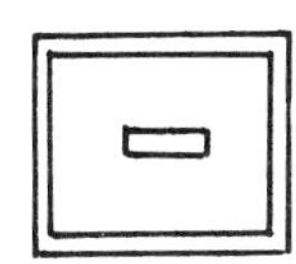

(Japan) House flag and symbol for Nichiro Gyogyo Kabushiki Kaisha (The Nichiro Fishing Co.) These floats are blue-green and vary from 2 to 12 inches in diameter. They were manufactured by Hakodate Glass K.K. and found on Vancouver Island in 1960.

2——

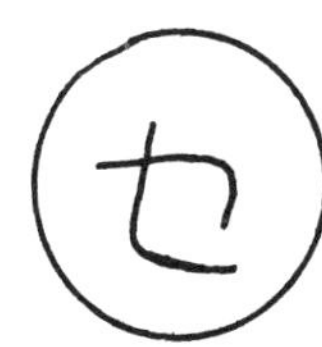

(Japan) Symbol is syllable *se* meaning "alphabet." These floats have been found in 2, 3, 6, 8, and 12-inch diameter along the Oregon, Washington, Vancouver Island, and Alaskan beaches. This marking was first reported in 1946 in Oregon.

3——

(Japan) Symbol is word *kita* meaning "north." It was used originally by Hokuyo (North Sea) Canned Crab Co. Floats are blue-green and have been found in 2, 4, 6, 8, 10, 12 and 14-inch diameter. First reported in 1946, they show up regularly in Oregon, Washington, Vancouver Island, Queen Charlotte Islands, and Alaska. Manufactured by Hokuyo Glass Co. of Aomori.

4——

(Japan) Symbol is words *kita chis* meaning "North Kurile." It is believed to be used by the Kita-Chisima Marine Products Co. The floats are blue and only two of these floats have been reported and these were picked up at Ocean Park, Washington; one was 6 inches, the other 8 inches in diameter.

5——
(Japan) Symbol is word *dai* meaning "big." Found in 1960 in Washington and Alaska. This symbol is found only on floats of 3-inch diameter.

6——
(Japan) Single float found at Long Beach, Washington, was blue-green and 3 inches in diameter.

7——
(Japan) House flag and symbol for Taiyo Gyogyo Kabushiki Kaisha (The Ocean Fishing Co.). This is considered the largest Japanese Fishing Company. Despite their extensive operations, this float is seldom found. Those recorded are one 3-inch blue-green at Copalis, Washington in 1936, and one 12-inch at Gearhart, Oregon, in 1960.

8——
(Japan) Symbol is word *shui* meaning "water." Used by Nemuro Suisan K.K. (Nemuro Fisheries Co.) which in 1963 was fishing in the Bering sea. Floats were manufactured by Hakodate Glass K.K. and Daiichi Glass K.K. A single 3-inch float was reported in the early 1950s in Washington.

9——
(Japan) Symbol is word *hu* meaning "seal." Found in blue and purple colors in Washington and Alaska. The purple float was 10-inch diameter. This symbol was first reported in 1958.

10——
(Japan or Korea) These floats were manufactured by Sakurai Glass K.K., and found first in Oregon prior to 1948 in 3-inch diameter. Also found at Vancouver Island, B. C., Wake Island, and Alaska in 1962. Symbol and user unknown.

11—

(Japan) Symbol is word *guchi* meaning "rivermouth." These floats are believed to be manufactured by Kawaguchi Glass K.K. They have been found on Vancouver Island, Queen Charlotte Islands, and in Alaska. Sizes are 2 and 3-inch diameter plus a roller type of 4-inch length. First reported in 1960 at Tofino, B. C.

12—

(Japan or Korea) These floats were manufactured by Sakurai Glass K.K. First reported at Tofino, B.C., in 3-inch size in 1960.

13—

(Japan) These floats were found first on Vancouver Island, B. C., in 1955, and more recently on Wake Island, the Queen Charlotte Islands and in Alaska. Sizes vary from 3, 4, 6, 8, through 10-inch diameter.

14—

(Japan) Symbol has words *tokui* and *mochimas* meaning "special" and "hold" or "keep." First reported at Long Beach, Washington, in 1935 and also noted in Kodiak, Alaska, in 1960. Only size reported was 16-inch diameter and blue in color. This float had a small hole in it which indicated thickness of glass to be 1/16 inch.

15—

(Japan or Korea) Part of symbol infers meaning of "heaven." One only found on Washington Coast —in 1958—of 10-inch diameter.

16—

(Japan) Symbol for syllable *ho*, with no meaning. A single float—blue in color and 2-inches in diameter — was found at the North jetty of the Columbia in 1964.

17——

(Japan) Symbol is initials of manufacturer. Made by Daiichi Glass K.K. *Ichi* means the "first." Found in 1960 on Vancouver Island, B.C., in 3-inch size.

18——

(Japan) This symbol apparently also belongs to Daiichi Glass K.K. These floats were found, starting in 1950, along the Washington Coast. Two and 3-inch sizes were picked up in 1963 in the Queen Charlotte Islands.

19——

(Japan) A single float of 3 inches diameter, bearing this symbol, was found in 1962 at Cape Alava, Washington.

20——

(Japan) Symbol is word *hi* meaning "sun." An 8-inch diameter blue-green float was found along the Washington Coast near Long Beach in 1950.

21——

(Japan) The single reported "find" with this symbol was by helicopter on the outer beach north of Tofino, B. C.

22——

(Japan) Symbol is word *toh* or *higashi*—both meaning "east." A single float with this symbol was found at Ocean Park, Washington, in 1956, a 3-inch size.

23——

(Japan) A 10-inch float containing this marking was found at Long Beach, Washington, in 1956.

24——

(Japan) Symbol is word *to* meaning "east." One 3-inch float containing this symbol was found at Long Beach, Washington, in 1963.

25——
(Japan) Symbol is word *sen* meaning "cork." Believed to be manufactured in the region south of Yokohama. Found regularly since 1960 in Oregon, Hawaii, Washington, and on Vancouver Island, B.C. All are blue-green and from 2, 3, 4, 6, 8, 10 through 14 inches in diameter. One was found at sea 800 miles north of Hawaii. They have probably been used recently in long-line tuna fishing operations.

26——
(Japan) A single 3-inch float was found with this marking on Vancouver Island, B. C., in 1960.

27——
(Japan) Symbol is said to mean "ten." One 3-inch float was found at Long Beach, Washington, in 1956.

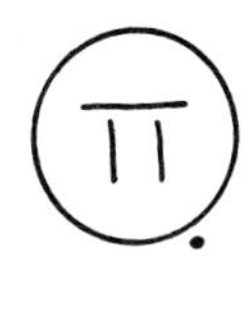

28——
(Japan or Korea) One 2-inch float was found at Tofino, B. C., in 1960; another at Wake Island in 1963.

29——
(Japan) User was Nemuro Kanzume K.K. (Nemuro Canning Co.) and manufacturer was Hakodate Glass K. K. A 3-inch size was found at Tofino, B. C., and Kodiak, Alaska, in 1960.

30——
(Japan) A single 3-inch float was found in 1956 at Long Beach, Washington.

31——
(Japan or Korea) This was found in 1960 at Ocean Park, Washington, in 3-inch size.

32——
(Japan) A single 3-inch float was found in 1960 at Ocean Park.

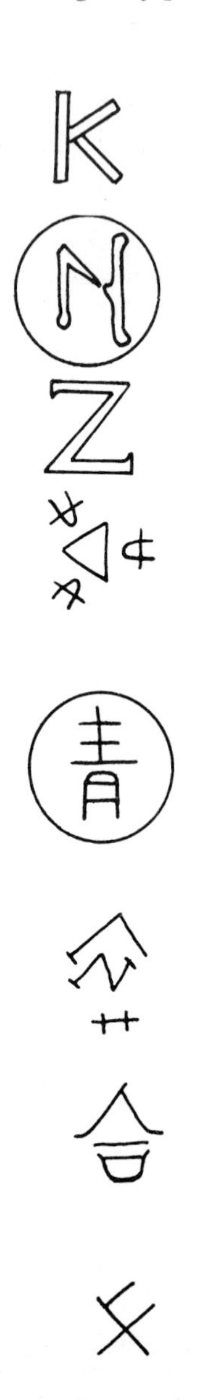

33—

(Japan or Korea) Found first on the Oregon Coast in 1946 and again, in 1960, on the Washington Coast. The 4-inch float was a Mason-jar blue color.

34—

(Japan) A 3-inch float was found at Long Beach, Washington, in 1956, with this marking.

35—

(Japan) A 3-incher with this symbol was found in 1963 at Ocean Park, Washington.

36—

(Japan) A 3-incher with this symbol was found in 1962 at Ocean Park.

37—

(Japan) Symbol is word *ching* meaning "green and blue." Findings were reported first at Long Beach, Washington, in 1954, and occasionally since, but just on Washington Coast beaches. Only large floats have contained this trademark—9, 10, 12 through 18-inch sizes.

38—

(Japan) A single 3-inch float was found at Ocean Park, Washington, in 1960.

39—

(Japan) Symbol is word *go* meaning "together." A single 8-inch float was found at Long Beach, Washington, in 1956.

40—

(Japan) This float was apparently made from a three-piece mold. The symbol was on both sides of the seal button. Two different people at Ocean Park, Washington, found the 3-inch blue-green floats in 1965. This is one of the newest markings reported.

41——
(Japan) A single 2-inch float was found at Tofino, B. C., in 1960. A second float was picked up on Wake Island in 1962.

42——
(Japan) A single 3-inch mold-made float was found at Ocean Park, Washington, in 1963, with this marking.

43——
(Japan) This unusual marking was found at Ocean Park, Washington, in 1963 in a 3-inch size.

44——
(Russia) Symbol is the hammer and sickle. User and manufacturer are presumed Russian. This marking was first reported in 1936 at Copalis, Washington, then in 1946 at Agate Beach, Oregon. The most recent find was in 1960 at Long Beach, Washington. All are the 3-inch size and mold made.

45——
(Japan) Symbol includes word *tani* meaning "valley." User was Nishitani Kaiun Kabushiki Kaisha of Otaru, Hokkaido. This company was active in the fishing districts of Hokkaido, Sakhalin, Kamchatka, and the Okhotsk Sea as early as 1933, and perhaps before then. A single mold-made float of 3-inch size was found at Long Beach, Washington, in 1960.

46——
(China) Symbol is *chu* meaning "master." A single 3 - inch float was found in 1960 at Long Beach.

47——
(China) Same symbol as No. 46 but with altered background. One 3-inch float bearing this marking was discovered at Long Beach, Washington, in 1960.

48——
(Russia) Symbol is a well-formed star. A single 3-inch float was found in 1960 at Long Beach.

49——
(Japan or Korea) Floats with this marking have been found in Washington and on Vancouver Island, B. C., since 1962. Glass color is light blue and floats have been picked up in 2, 3, and 10-inch diameter sizes.

50——
(Japan or Korea) Floats bearing this symbol have been found twice on the Washington Coast since 1960. Sizes are 2 and 3 inches in diameter. Glass is light blue.

51——
(Japan) A single 2-inch float was found at Tofino, B. C., in 1960.

52——
(Japan) A single 3-inch float with this symbol was found at Long Beach, Washington, in 1958.

53——
A clear, white glass float with this symbol was found at Kodiak, Alaska, in 1962. Not identified.

54——
(Japan) Symbol includes the word *sa* meaning "unknown." Manufacturer might have been Sasa Glass Co., which was in business in 1949. A single blue-colored 4-inch float was found at Ocean City, Washington, in 1957.

55——
(Japan) Symbol includes Nichiro house flag and the word *sa* meaning "unknown." User might have been Nichiro Fishing Co. Manufacturer might have been Sasa Glass Co. A single 3-inch float was found at Ocean Park, Washington, in 1962.

56—
(Japan) Symbol contains the word *sa* meaning "unknown." Maker might have been Sasa Glass Co. The mark was first reported in Kodiak, Alaska, in 1960—then at Ocean Park, Washington, in 1962. Size was 3 inches in diameter.

57—
(Japan) A single 3-inch float with this symbol was found at Ocean City, Washington, in 1957.

58—
(Japan) A single 3-inch float was found in 1963, at Ocean Park, Washington.

59—
(Japan) Symbol is word *betsu* meaning "different." A 14-inch blue-green float using this mark was found in 1950 at Long Beach, Washington.

60—
(Japan) Symbol is varied form of word *betsu* meaning "different." A single green-color float was found in Kodiak, Alaska, in 1960.

61—
(Japan) Symbol is another form of the word *betsu* meaning "different." In 1960, a green float with this mark was picked up near Kodiak, Alaska.

62—
(Japan or Korea) A single 3-inch float bearing this insignia was found in 1963 off the Washington Coast.

63—
(Japan or Korea) This is a three-piece mold found in the 3-inch-diameter size along Oregon and Washington Coasts; also on the Queen Charlottes and Vancouver and Wake Islands.

64——
(Japan or Korea) A three-piece mold found widely since 1960.

65——
(Japan or Korea) A three-piece mold found widely since 1960.

66——
(Japan or Korea) A three-piece mold found widely since 1960.

67——
(Japan) A single 3-inch float bearing this mark was picked up at Ocean Park, Washington, in 1960.

68——
(Japan or Korea) A float with this mark was found off Vancouver Island in 1962.

69——
(Japan) This mark first showed in Oregon in 1946, then in Alaska in 1960.

70——
(Japan) A 3-inch size was found in Bristol Bay, Alaska, in 1964. Float was the usual blue-green color with a dark-blue sealing button.

71——
(Japan) A 3-inch size was found in Bristol Bay, Alaska, in 1964. Float was blue-green in color with a dark-blue sealing button.

72——
(Japan) A 3-inch size was found in Bristol Bay, Alaska, in 1964. Float was blue-green in color with a dark-blue sealing button.

73——
(Japan) A 3-inch size was found in Bristol Bay, Alaska, in 1964. Float was blue-green in color with a dark-blue sealing button.

74——
(United States) A single, machine-made clear 7-inch float with this imprint was found at Kalaloch, Washington in 1959.

75——
(Norway) A trademark imprint found on floats used in North Atlantic waters. Manufactured by Flesland Glasverk of Bergen, Norway. The floats are amber and blue-green, 4 inches in diameter.

76——
(Great Britain) A trademark imprint. These were known to exist by 1948. They are clear, machine-made 6-inch-diameter floats bearing the brand opposite the hand-sealed end.

MADE
IN
JAPAN

77——
(Japan) An export marking imprint. These floats, from 6 inches through 12 inches, have been widely found since first seen in Oregon in 1948. They were sold to American fishermen for the soupfin shark trade in 1939, when imported German-made floats were no longer available.

NW

78——
(United States) A trademark imprint of Northwestern Glass Co. of Seattle, Washington. Glass floats from 3½ through 6-inch size of flint and amber color have been found widely since 1948. They are machine made.

79——
(United States) The imprint of a Russian Company. The user of these American-made machine-type floats was Krabotrest, a crab-fishing company of Vladivostok, U.S.S.R. About 690,000 of these floats were manufactured by Northwestern Glass Co. on lend-lease in 1943. All were delivered to a Russian ship in Seattle. These floats are clear white and 3½-inch size. One was reported at Long Beach, Washington, in 1958.

80—

(United States) An imprint on an experimental production of small amber bottles with sealed neck. Few, if any, are believed to exist since they were tried in 1941.

81—

(United States) A trademark imprint of original handblown floats made by Northwestern Glass between 1933-36. These were clear glass and 8-inch size. About 40,000 were manufactured. Since most of these were used in Alaskan waters, relatively few are found on Oregon and Washington beaches.

82—

(United States) A single machine-made, clear 5-inch float with this imprint was found at Ocean Park, Washington, in 1963. The float had a small sealing button only ½-inch in diameter.

83—

(Germany) A trademark imprint. American Pacific Coast fishermen used these imported high-quality glass floats in the early 1930s in off-shore and Alaskan fishing. They were a pale-green color in both 3 and 8-inch sizes. First reported at Cape Alava, Washington, in 1935, in Oregon in 1948, and Alaska in 1960.

84—

(United States) Trademark for Owens-Illinois. In about 1943, this glass factory at Oakland, California, began turning out machine-produced, clear glass floats of 6-inch diameter. They have occasionally been found on Washington beaches since 1960.

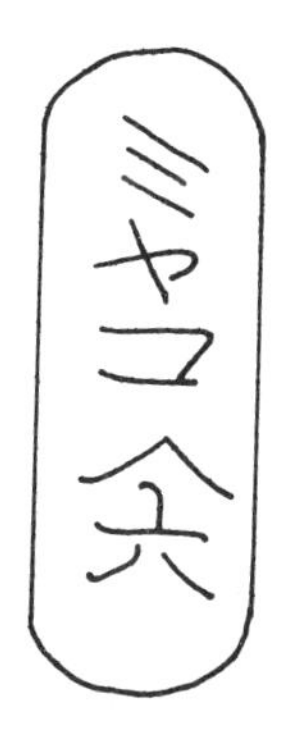

85——

(Japan) This symbol is believed to be the name of a fishing fleet. It was found in Oregon, in the 1940s, on a 4-inch-long roller-type float.

86——

(Japan) This symbol includes the words *oo bune watashi*, meaning "large boat across river." The upper part of the symbol is believed to be a family crest. The float was found on Wake Island in the early 1960s.

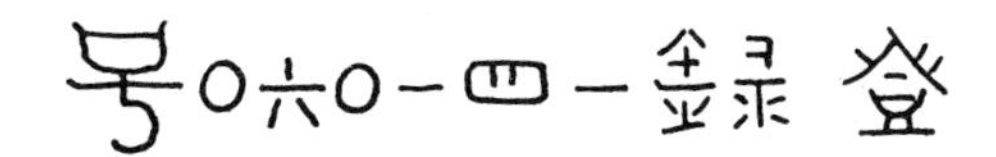

87——

(Japan) The symbol reads "Patent Applied for, No. 141060." This marking in large characters was on a 12-inch-diameter mold-made float containing a hollow glass tube. It was reported picked up near Agate Beach, Oregon, in 1948.

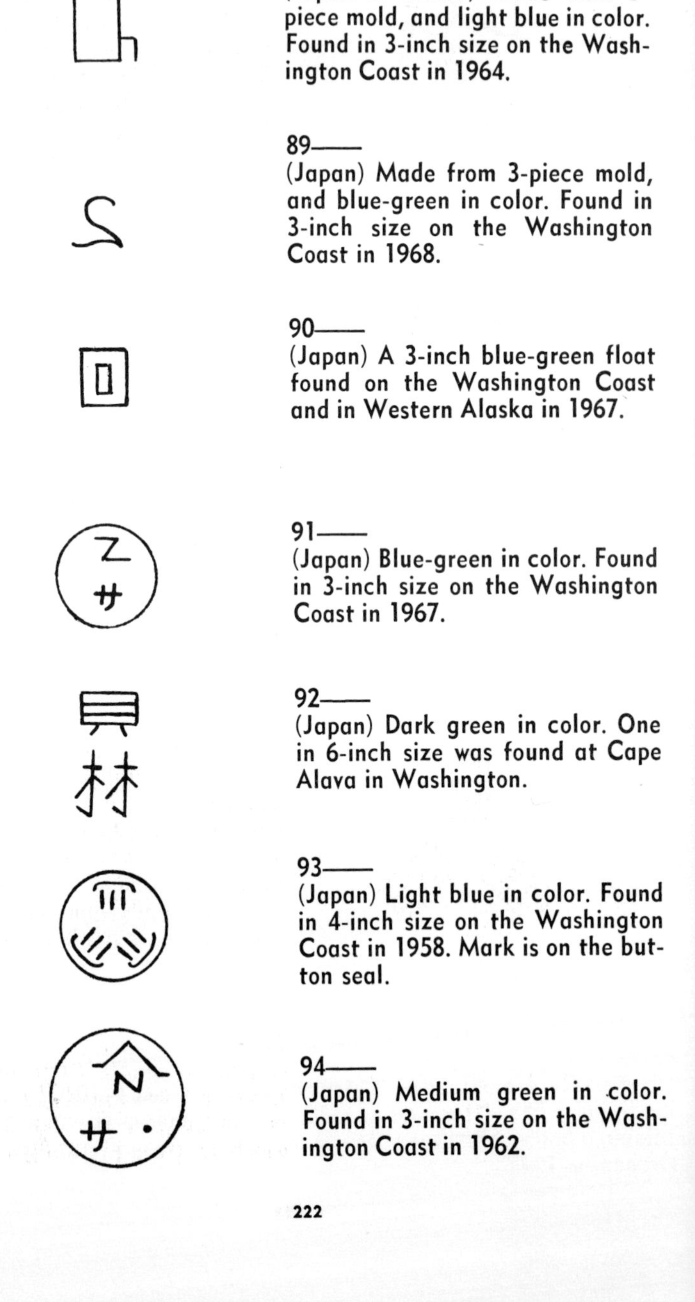

88—
(Japan or Korea) Made from 3-piece mold, and light blue in color. Found in 3-inch size on the Washington Coast in 1964.

89—
(Japan) Made from 3-piece mold, and blue-green in color. Found in 3-inch size on the Washington Coast in 1968.

90—
(Japan) A 3-inch blue-green float found on the Washington Coast and in Western Alaska in 1967.

91—
(Japan) Blue-green in color. Found in 3-inch size on the Washington Coast in 1967.

92—
(Japan) Dark green in color. One in 6-inch size was found at Cape Alava in Washington.

93—
(Japan) Light blue in color. Found in 4-inch size on the Washington Coast in 1958. Mark is on the button seal.

94—
(Japan) Medium green in color. Found in 3-inch size on the Washington Coast in 1962.

95——
(Japan) This trademark also belongs to Daiichi Glass K.K. (See No. 17.) Floats with this mark have been found since 1960 along the Oregon and Washington beaches. These floats are found only in the 2 and 3-inch sizes.

96——
(Japan) This trademark is another variation of the Daiichi Glass K.K. production. These floats have been found on Oregon and Washington beaches since 1960 in the 3-inch size.

97——
(Japan) Believed to be another form of the trademark for Kawaguchi Glass K.K. (See No. 11.) These have been found along Oregon and Washington beaches since 1960 in the 3-inch size.

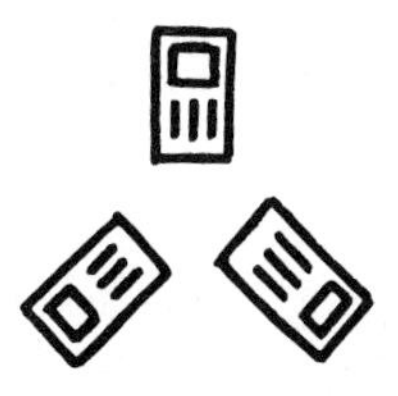

98——
(Japan) This is believed to be still another trademark of Kawaguchi Glass K.K. Floats with this imprint have been found along Oregon and Washington beaches, also in Alaska and Hawaii, since 1960. The color is usually light blue, and it often occurs in smaller size rollers.

99——
(Japan) A single float with this trademark was found on the Washington Coast in the 3-inch size. It was blue-green in color.

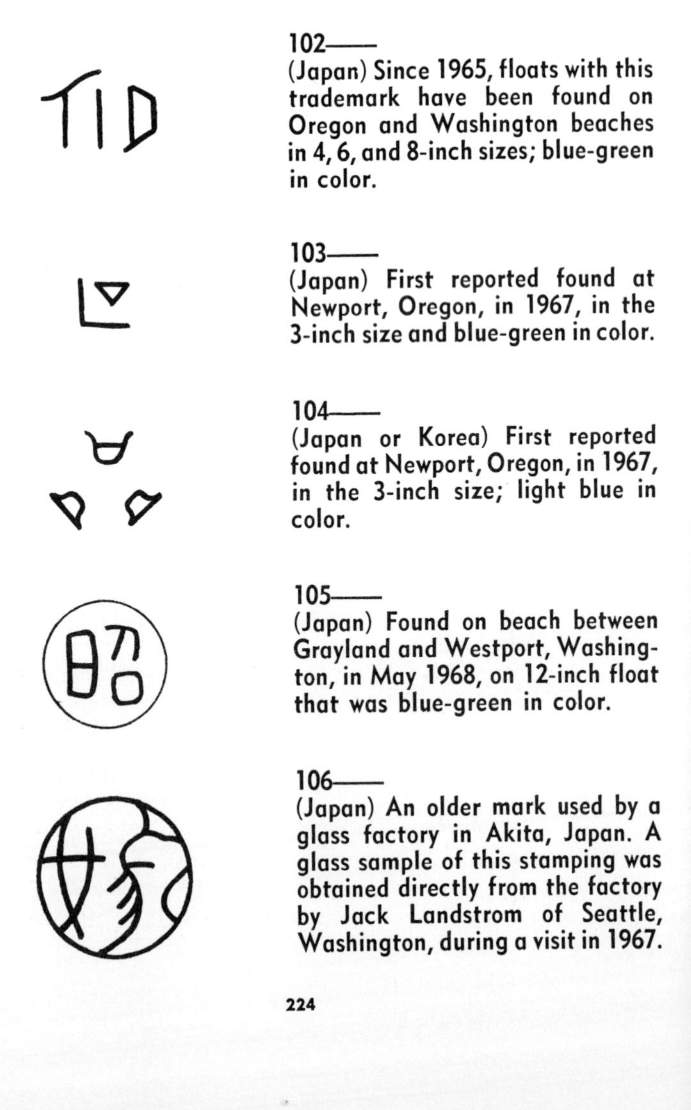

100—
(Korea) Found on the Oregon Coast in 1964; 3-inch size, and 3-piece mold made.

101—
(Korea) Found near Oysterville, Washington, in 1967; 3-inch size, and 3-piece mold made.

102—
(Japan) Since 1965, floats with this trademark have been found on Oregon and Washington beaches in 4, 6, and 8-inch sizes; blue-green in color.

103—
(Japan) First reported found at Newport, Oregon, in 1967, in the 3-inch size and blue-green in color.

104—
(Japan or Korea) First reported found at Newport, Oregon, in 1967, in the 3-inch size; light blue in color.

105—
(Japan) Found on beach between Grayland and Westport, Washington, in May 1968, on 12-inch float that was blue-green in color.

106—
(Japan) An older mark used by a glass factory in Akita, Japan. A glass sample of this stamping was obtained directly from the factory by Jack Landstrom of Seattle, Washington, during a visit in 1967.

107——
(Japan) A 3-inch size was found on Vancouver Island in 1965. Float was blue-green with a dark blue sealing button.

108——
(Japan) A 2-inch size was found on Vancouver Island in 1965. Float was green in color.

109——
(Japan) This symbol was found on a 5-inch roller on Vancouver Island in 1965.

110——
(Japan) A 3-inch size was found on Vancouver Island in 1965. Float was light blue-green in color.

111——
(Japan) A 3-inch size was found on Vancouver Island in 1965. Float was blue-green in color.

112——
(Japan) A 13-inch float containing this marking was found in Bristol Bay, Alaska, in 1968. Float was 3-piece-mold constructed, aqua-green in color. Netting was ½-inch cotton rope.

113——
(Japan) A 10-inch float containing this marking was picked up on the ocean beach in the vicinity of Lake Ozette, Washington, in 1961. Mark was on separate button; float was blue-green in color.

114—
(Japan) A 3-inch float, green in color, was found in the Bristol Bay, Alaska, region in 1966.

115—
(Japan) A 3-inch float, dark red in color, was found near Seward, Alaska, in 1967.

116—
(Japan) A 3-inch float with this marking was picked up near Kodiak, Alaska, in 1967.

117—
(Japan) A 3-inch float with this marking was found on the ocean beach in the vicinity of Lake Ozette, Washington, in 1958.

118—
(Japan) Floats with this marking have been found in Oregon, Washington, and Alaska as early as 1963 in the 3-inch size.

119—
(Japan) This imprint is the house mark of a glass factory in Akita Prefecture, Japan. It stands for *Takahashi,* the name of the owner at that time.

120—
(Japan) A 6-inch float containing this marking was found at Yachats, Oregon, in 1962. The float was heavily built and purple in color.

LT

121——
(Japan or England) A 6-inch float—purple in color and completely frosted on the exterior—was found by Nancy Crane in the Gulf of Mexico in 1968, on South Padre Island, Texas. Others, white and faint pink, have been found in the British West Indies.

122——
(Unknown) Several floats with this imprint in different colors were found by Grethe Seim on Grand Turk Island in the British West Indies, in 1967.

123——
(Japan) An 8-inch roller float bearing this marking and with a mold line was found at Neah Bay, Washington, during the winter of 1967-68. It was reported by Lt. Thomas J. Spittler of the Air Force Station. The top two characters, *Mi-ya-ko,* mean "Capital City"; the bottom three, *To-ku-e,* "Special Picture." The central, circled character is unknown. Manufacturer and user are also unknown.

124——
(Japan) This unusual 4-inch roller-type float was found in Southeast Alaska in 1962. It is believed to have been manufactured about 1930. The marking appears in 3 parts. The two upper groups, *Ka-mano-saki,* refer to a location or city. The middle group represents a brand name or symbol. The two lower groups, *Chu-kichi,* are the name of a person.

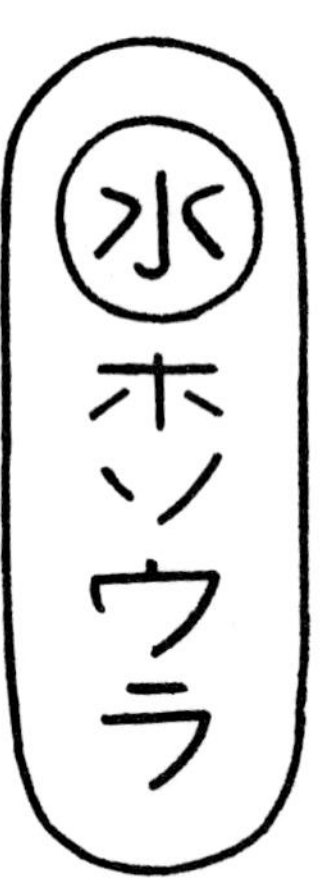

125——
(Japan) This very unusual 6-inch roller-type float was found in Southeast Alaska in 1958. It had longitudinal mold marks the entire length. The glass was clear, colorless, and free of bubbles. The markings read, *Maru mizu ho-so-u-ra,* meaning "Round water slender back." This float is in the collection of Dr. Leland Paddock of Sequim, Washington.

126——
(Japan) A 5-inch float bearing this marking was found at Tofino, Vancouver Island, in 1964. The float was blue in color.

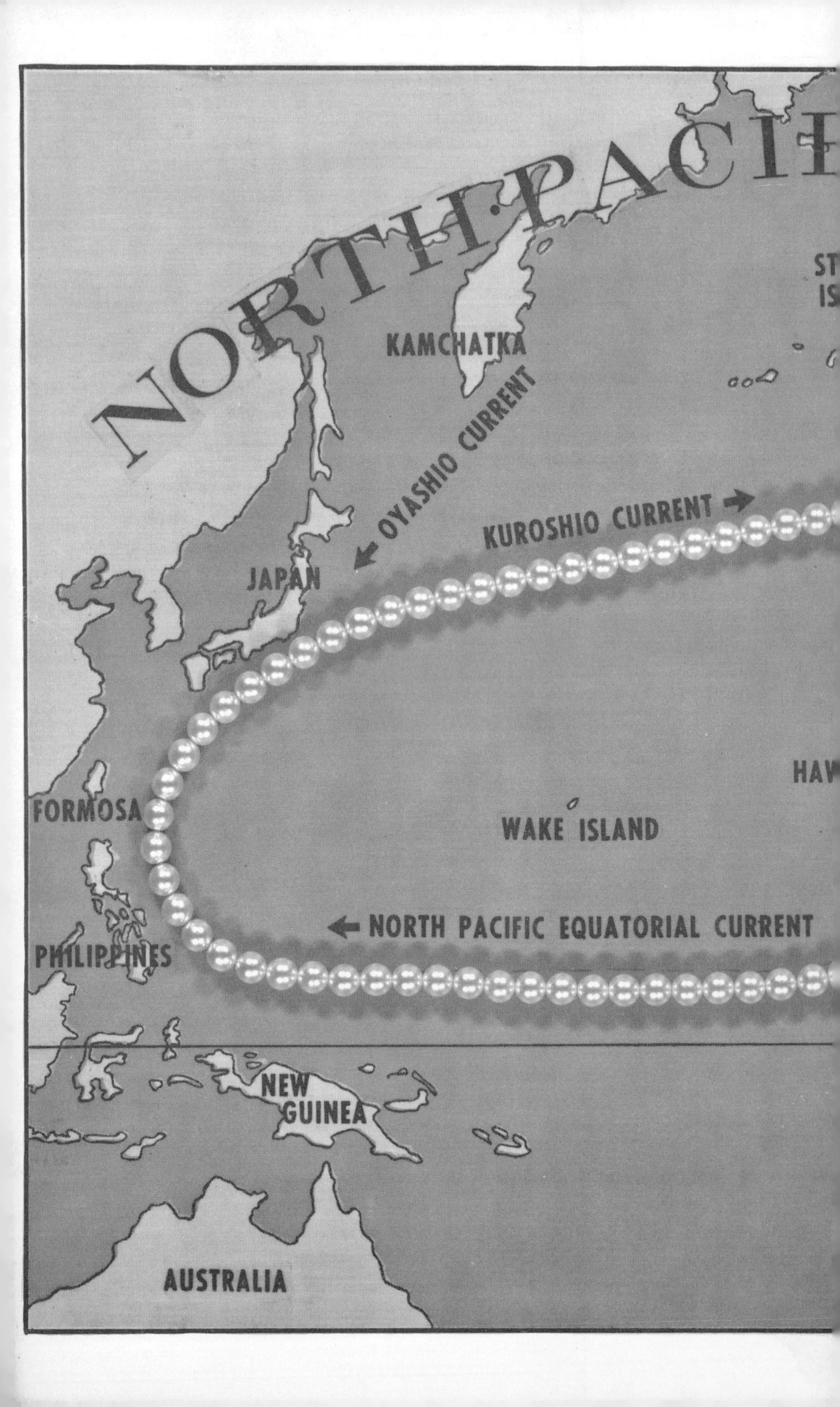

NORTH
KAMCHATKA
OYASHIO CURRENT
KUROSHIO CURRENT
JAPAN
FORMOSA
WAKE ISLAND
NORTH PACIFIC EQUATORIAL CURRENT
PHILIPPINES
NEW GUINEA
AUSTRALIA